Logic grid puzzles, also called Einstein puzzles or Zebra puzzles, provide readers with a set of clues intended to help them establish the links between items with different characteristics. Each puzzle has only one solution, which can be discovered through the process of logical deduction. The characteristics are listed on logic grids, where known information will be collected and false connections will be eliminated until only the correct answers remain.

This book contains an example 2x2 logic grid with a detailed solution and 50 3x3 puzzles, ranging from easy to difficult. Solutions can be found at the end of the book.

All puzzles in this book were solved by a human expert to ensure have sufficient clues to help you find a solution, including the hard puzzles at the end of the book – however, every human can make mistakes. If you spot a mistake, please let us know by mailing fpp.puzzles@error300.com. We'll also be glad to hear any comments/requests you might have – we are making these books for you and we'd really love you to enjoy them.

Our puzzles might be harder than the logic grids you're used to, but every single one is solvable – and a full solution for each puzzle is available under the QR code provided by the puzzle name. Obviously, those solutions were also done by humans and a mistake could happen, but before calling out a puzzle has insufficient clues, please make sure you didn't miss something (yes, there is a difference between "somewhere right of" and "directly right of"). If some step in the solution seems unclear, feel free to reach out to us or to try our books on solving logic grid puzzles:

Basic Example

Conditions

Five children are celebrating birthdays in consecutive months.

1. Francine's birthday is in August.
2. Jason's birthday is in October, and he's a year older than Celeste.
3. The child turning 14 wasn't born in June.
4. The children celebrating birthdays in June and August were born in consecutive years, in no specific order.
5. Jason is one year younger than the person born in July.
6. Peter is one year older than Emily.

		Age					Month				
The Birthdays		11	12	13	14	15	June	July	August	September	October
Name	Emily										
	Peter										
	Jason										
	Francine										
	Celeste										
Month	June										
	July										
	August										
	September										
	October										

Initial Elimination

We can begin by marking the grid with any information given to us in the clues.

In this case, Clues 1 and 2 give us the birthdays of Jason and Francine. Since each child has only one possible birth month, we can eliminate all other months for Jason and Francine. Moreover, because each child's birthday is a different month, we can also eliminate all other children's names for August and October.

Clue 3 allows us to eliminate another possibility: We know the child turning 14 was not born in June, so we should mark that off as well.

		Age					Month				
The Birthdays		11	12	13	14	15	June	July	August	September	October
Name	Emily								X		X
	Peter								X		X
	Jason						X	X	X	X	✓
	Francine						X	X	✓	X	X
	Celeste								X		X
Month	June				X						
	July										
	August										
	September										
	October										

Deduction

We know from Clue 6 that Peter is one year older than Emily, so Peter cannot be the youngest child, and Emily cannot be the oldest. Hence, we eliminate the youngest age for Peter, and the oldest age for Emily. Similarly (from Clue 2), Jason cannot be the youngest and Celeste cannot be the oldest. Clue 5 also uses a similar line of logic, allowing us to deduce that Jason cannot be the oldest and the person with the July birthday cannot be the youngest. Also, we know that Jason is older than Celeste but younger than the person born in July, which means Celeste cannot have been born in July.

Clue 4 tells us the children born in June and August were born in consecutive years. We know Francine was born in August, from Clue 1. If she was 15 years old, it would imply the person born in June is 14. However, Clue 3 clearly states that the child turning 14 wasn't born in June, which means it's impossible for Francine to be 15. Therefore, we can deduce that the only option for age 15 is Peter.

		Age					Month				
	The Birthdays	11	12	13	14	15	June	July	August	September	October
Name	Emily					X		X			X
	Peter	X				✓			X		X
	Jason	X			X	X	X	X	X	X	✓
	Francine					X	X	X	✓	X	X
	Celeste					X		X	X		X
Month	June				X						
	July	X									
	August										
	September										
	October										

With each age we confirm, we eliminate the other possibilities within the corresponding column and row, allowing us to eliminate the other ages for Peter.

Since Emily is one year younger than Peter, we now know Emily's age as well: age 14. Consequently, Clue 3 tells us Emily couldn't have been born in June.

Additionally, if Peter were born in June, Emily and Francine would be the same age, because Clue 4 tells us Francine and Peter were born in consecutive years, and we've already determined Peter is the oldest. It's impossible for two children to share the same age; thus, Peter couldn't have been born in June. That leaves us with only one possibility for the child born in June: Celeste.

We know that Jason is one year younger than the person born in July from Clue 5. This eliminates July as a possibility for Peter because, if Peter were born in July, both Jason and Emily would be one year younger than Peter—the same age. As a result, the only possibility for July is Emily, which leaves September as the only remaining option for Peter.

	The Birthdays	11	12	13	14	15	June	July	August	September	October
Name	Emily	X	X	X	✓	X	X	✓	X	X	X
	Peter	X	X	X	X	✓	X	X	X	✓	X
	Jason	X			X	X	X	X	X	X	✓
	Francine			X	X		X	X	✓	X	X
	Celeste						✓	X	X	X	X
Month	June				X						
	July		X								
	August										
	September										
	October										

All we need to do now is establish the remaining ages.

From Clue 5 we know Jason is one year younger than the person born in July, who happens to be Emily – and we already know her age. This would mean Jason is 13.

Also, from Clue 2, Jason happens to be a year older than Celeste – so Celeste must be 12, leaving us with only one option for Francine – 11.

And the puzzle is solved!

	The Birthdays	11	12	13	14	15	June	July	August	September	October
			Age					Month			
Name	Emily	✗	✗	✗	✓	✗	✗	✓	✗	✗	✗
	Peter	✗	✗	✗	✗	✓	✗	✗	✗	✓	✗
	Jason	✗	✗	✓	✗	✗	✗	✗	✗	✗	✓
	Francine	✓	✗	✗	✗	✗	✗	✗	✓	✗	✗
	Celeste	✗	✓	✗	✗	✗	✓	✗	✗	✗	✗
Month	June			✗							
	July	✗									
	August										
	September										
	October										

We'd like to encourage you to join our newsletter and as a thank you for doing so we'd like to offer you two free digital books:

They can be downloaded here:

As a side little bonus we've also added a printable empty 3x3 grid sheet you can use as backup.

Puzzles

Five Ships

There are five merchant ships preparing to depart from five different docks. The docks are numbered in increasing order, 1 through 5, with Dock 1 being the leftmost one. Each ship has a different captain and is transporting a different product.

What product is the Sea Queen carrying, and what is the name of its captain?

1. The Sea Queen is in Dock 2, somewhere left of the Mermaid.
2. Four things we know about Captain Millman's ship.
   a. First, it is docked somewhere to the left of Croydon's ship.
   b. Second, it is somewhere to the left of the Neptune (although we don't know the captain of the Neptune).
   c. Third, it is in the dock adjacent to the silk merchant's ship.
   d. Fourth, it is docked somewhere to the right of the Mermaid.
3. The spice merchant's ship is docked somewhere to the left of the ship captained by Murdoch.
4. The ships selling jewels and tobacco are in adjacent docks.
5. There is one dock between where the Fairy is docked and where the jewel merchant's ship is docked, in some order.
6. The ship captained by Lennox is docked somewhere left of the ship selling spices.

12

Five Ships

		Dock					Captain					Merchandise				
Five Ships		1	2	3	4	5	Croydon	Harris	Lennox	Millman	Murdoch	Gold	Jewels	Silk	Spices	Tobacco
Name	Fairy															
	Mermaid															
	Neptune															
	Sea Queen															
	Valentine															
Merchandise	Gold															
	Jewels															
	Silk															
	Spices															
	Tobacco															
Captain	Croydon															
	Harris															
	Lennox															
	Millman															
	Murdoch															

Name	Dock	Captain	Merchandise

Restaurant Recommendations

There's no shortage of places to try if you want to eat out in the city! Five new restaurants have opened recently, all offering a different type of cuisine. Each of them is painted a different color and has received a rating from 3 to 5 stars from the local food critic, each having a half star more than the last.

What type of food can you find at Happy Tummy and how many stars is it rated?

1. Hungry No More is rated somewhat lower than the blue restaurant, but somewhat higher than the white one.
2. Three things we know about the gray restaurant.
    a. First, it has 0.5 stars more or 0.5 stars less than the Thai restaurant.
    b. Second, it is rated somewhat higher than the red restaurant.
    c. Third, it has either 0.5 stars less or 0.5 stars more than the white restaurant.
3. The red restaurant has 0.5 stars less than Nice Food Diner, but is rated somewhat higher than the yellow restaurant.
4. Four things we know about the Indian restaurant.
    a. First, it has a somewhat higher rating than the Italian restaurant.
    b. Second, it's rated higher than Delicious (although it's not clear what type of food Delicious offers).
    c. Third, it has either 1.5 stars more or 1.5 stars less than the blue restaurant.
    d. Fourth, it has a somewhat lower rating than Yum Yum.
5. The Chinese restaurant has either 1 star more or 1 star less than the blue restaurant.

Restaurant Recommendations

	Restaurant Recommendations	Stars					Cuisine					Color				
		3	3.5	4	4.5	5	Chinese	French	Indian	Italian	Thai	Blue	Gray	Red	White	Yellow
Name	Delicious															
	Happy Tummy															
	Hungry No More															
	Nice Food Diner															
	Yum Yum															
Color	Blue															
	Gray															
	Red															
	White															
	Yellow															
Type	Chinese															
	French															
	Indian															
	Italian															
	Thai															

Name	Stars	Cuisine	Color

The Christmas Ornaments Riddle

Each year, this family has a unique and cherished tradition: from the 19th to the 23rd of December, each of the children of the house are in charge of putting one type of ornament on the Christmas tree, so it is fully decorated by Christmas Eve. Each type of ornament has a different shape and color.

What do the ornaments Avery is in charge of look like, both in shape and in color?

1. Mateo will decorate the tree either the day before or the day after the child in charge of the gold ornaments decorates.
2. Mateo will definitely decorate sometime before Elijah and sometime after Paisley, though.
3. The white ornaments will be on the tree before Paisley decorates.
4. The reindeer-shaped ornaments will not be put on the tree on the 20th.
5. Stella will decorate the tree sometime after the child in charge of the green ornaments decorates.
6. Also, the ornaments Stella's in charge of and the Santa-shaped ornaments will be put on the tree on consecutive days, in some order.
7. The star-shaped ornaments will be placed on the tree sometime before the bell-shaped ones, but sometime after the green ornaments.
8. The red ornaments will be placed on the tree sometime after Elijah has his turn to decorate.

The Christmas Ornaments Riddle

The Christmas Ornaments Riddle		Date					Ornament					Color				
		19	20	21	22	23	Angel	Bell	Reindeer	Santa	Star	Gold	Green	Red	Silver	White
Child	Avery															
	Elijah															
	Mateo															
	Paisley															
	Stella															
Color	Gold															
	Green															
	Red															
	Silver															
	White															
Ornament	Angel															
	Bell															
	Reindeer															
	Santa															
	Star															

Child	Date	Ornament	Color

Custom Dice

Leon did a test 3D print of a set of custom dice, with numbers on some faces and designs on the other faces. He has designed five different dice, each with a feline on two faces, a shape on two faces, and a letter on two faces. Each of the five dice have different animals, shapes, and letters on them. Leon has lined the test prints up in front of him, we'll assume the die on the far left is Die 1, and number them in increasing order from left to right.

What shape and letter are printed on the die with the lion design?

1. The die with panthers on some of its faces and the die with triangles on some of its faces are sitting next to each other, in some order.
2. Two things we know about the die with leopard faces.
    a. First, it has a letter "D" on some of its other faces.
    b. Second, the leopard die is somewhere left of the die with squares on some of its faces.
3. Two things we know about the die with cheetahs on some of its faces.
    a. First, it is right of the die with the letter "A" on some of its faces, with one die between them.
    b. Second, it is somewhere right of the die with hexagons on some of its faces (though we don't know what other symbols are on the dice with hexagon faces).
4. Four things we know about the die with lions on some of its faces.
    a. First, it is somewhere to the right of the die with circles on some of its faces.
    b. Second, it is directly to the right of the die with pentagons on some of its faces.
    c. Third, it is somewhere to the right of the die with squares on some of its faces.
    d. Fourth, the die with lions on some of its faces is next to the die with the letter "Z" on some of its faces, in some order.
5. The die with the letter "F" on some of its faces is somewhere to the right of the die with "H" on some of its faces.
6. There is one die between the die with circles on some of its faces and the die with squares on some of its faces, in some order.

Custom Dice

Custom Dice		Order					Shape					Letter				
		1	2	3	4	5	Circle	Hexagon	Pentagon	Square	Triangle	A	D	F	H	Z
Feline	Cheetah															
	Leopard															
	Lion															
	Panther															
	Tiger															
Letter	A															
	D															
	F															
	H															
	Z															
Shape	Circle															
	Hexagon															
	Pentagon															
	Square															
	Triangle															

Feline	Order	Shape	Letter

Alien Planets

Five habitable planets have been discovered in a faraway galaxy. Each planet is the same direction from Earth, but a different amount of parsecs away, with 10 parsec being closest and 50 parsecs being farthest. Each planet has a different-colored sun, a different name, and possesses alien inhabitants with different personalities.

What personality do the inhabitants of Thaket have, and what color is their sun?

1. Two things we know about the planet with a red sun.
    a. First, there are 30 parsecs between this planet and the planet with a purple sun.
    b. Second, the planet with a red sun is somewhat closer to Earth than the planet with selfish inhabitants.
2. Two things we know about Reshyu.
    a. First, it is either 10 parsecs closer to Earth or 10 parsecs farther from Earth than the planet with shy inhabitants.
    b. Second, it is some distance farther away from Earth than Gamush.
3. Three things we know about the planet with suspicious inhabitants.
    a. First, it is somewhat closer to Earth than the planet with friendly inhabitants.
    b. Second, it is 10 parsecs closer to or 10 parsecs farther from Earth than the planet with a gray sun.
    c. Third, it is somewhat farther from Earth than the planet with the blue sun.
4. Two things we know about Ardyk.
    a. First, there is one planet between Ardyk and Dongyhuk.
    b. Second, it is somewhat closer to Earth than the planet with a gray sun.
5. Two things we know about the planet with a purple sun.
    a. First, it is some distance farther from Earth than Donghyuk.
    b. Second, it is not the farthest planet from Earth, out of the five.
6. Dongyhuk is somewhat farther from Earth than the planet with a blue sun, but somewhat closer to Earth than the planet with a green sun.

Alien Planets

	Parsecs					Personality					Sun color				
Alien Planets	10	20	30	40	50	Aggressive	Friendly	Selfish	Shy	Suspicious	Blue	Gray	Green	Purple	Red
Name Ardyk															
Dongyhuk															
Gamush															
Reshyu															
Thaket															
Sun color Blue															
Gray															
Green															
Purple															
Red															
Personality Aggressive															
Friendly															
Selfish															
Shy															
Suspicious															

Name	Parsecs	Personality	Sun color

Forest Findings

Lucien loves hiking in the forest and collecting things he finds there. He's stored his best-preserved specimens in five different boxes in front of him, starting with Box 1 on the far left and continuing to the right in numerical order. Each box is made from a different type of wood (retrieved from the forest he found the object in), has a different find inside, and is opened by a different-colored key.

What box does the gold key open, what wood is that box made of, and what does it contain?

1. Two things we know about the box made of ash wood.
   a. First, it is somewhere to the right of the box opened by the tan key.
   b. Second, it is somewhere to the left of the aspen box.
2. We know two other things about the aspen box.
   a. First, it's next to the box containing flower petals, in some order.
   b. Second, it's next to the box opened by the mint-colored key, in some order (we don't know the contents of this box though).
3. Four things we know about the box opened by the silver key.
   a. First, it is somewhere to the left of the box containing pine cones.
   b. Second, there are two boxes between the box with the silver key and the box containing flower petals, in some order.
   c. Third, there is one box between the box that opens with the silver key and the box that opens with the lilac key, in some order.
   d. Fourth, the silver-key box is somewhere left of the box containing feathers.
4. The box containing leaves and the hickory box are next to each other, in some order.
5. The alder box is somewhere to the right of the box that contains pine cones, with one other box between them.
6. The box containing stones is somewhere to the right of the box containing feathers.

Forest Findings

	Forest Findings	Box					Wood					Key color				
		1	2	3	4	5	Alder	Ash	Aspen	Cherry	Hickory	Gold	Lilac	Mint	Silver	Tan
Contents	Feathers															
	Flower petals															
	Leaves															
	Pine cones															
	Stones															
Key color	Gold															
	Lilac															
	Mint															
	Silver															
	Tan															
Wood	Alder															
	Ash															
	Aspen															
	Cherry															
	Hickory															

Contents	Box	Wood	Key color

A Week's Worth of Meals

Daisy is a very organized person and always likes being on top of things. Today, she decided to meal prep for the next work week. This includes preparing a sandwich, juice, and dessert to take to work each day, Monday through Friday.

What sandwich, dessert, and juice will Daisy bring on Monday?

1. Daisy is having a cupcake for dessert either the day before or the day after she brings pineapple juice.
2. There is one day between the day Daisy is having orange juice and the day she's having the granola bar, in some order.
3. Daisy is having fruit for dessert sometime later than she's having a cheese sandwich.
4. She's also having the cheese sandwich sometime before she'll eat a granola bar.
5. There are two days between the day Daisy has a ham sandwich and the day she's having yogurt for dessert, in some order.
6. Yogurt day is also either the day before or the day after the day she'll drink apple juice.
7. Three things we know about the day Daisy will bring a tomato sandwich.
    a. First, Daisy will enjoy this sandwich sometime later than she'll have a granola bar, but sometime earlier than when she'll bring apple juice.
    b. Second, there is one day between the day Daisy has a tomato sandwich and the day she's having peach juice, in some order.
    c. Third, she's having the tomato sandwich sometime later than the egg salad sandwich.

A Week's Worth of Meals

A Week's Worth of Meals		Day					Dessert					Juice				
		Monday	Tuesday	Wednesday	Thursday	Friday	Cookie	Cupcake	Fruit	Granola bar	Yogurt	Apple	Lemonade	Orange	Peach	Pineapple
Sandwich	Ham															
	Cheese															
	Egg salad															
	Tomato															
	Tuna															
Juice	Apple															
	Lemonade															
	Orange															
	Peach															
	Pineapple															
Dessert	Cookie															
	Cupcake															
	Fruit															
	Granola bar															
	Yogurt															

Sandwich	Day	Dessert	Juice

Book Day

It was book day at school today, and each student brought a book they wanted to share with the class. We will focus on the first row of five desks, in ascending numerical order, with Desk 1 on the far left. Each student has brought a different book, of a different genre.

What child brought "The Best of Jokes" and what is the genre of that book?

1. The child in the second desk did not bring an action book.
2. Vince and Cam sit next to each other.
3. Vince also sits somewhere right of the student who brought the action book.
4. As for Cam, there is one desk between his desk and the desk of the child who brought the horror novel, in some order.
5. Arnie sits somewhere left of the student who brought "Finding Theo."
6. Also, there is one desk between Arnie's desk and the desk of the kid who brought the horror novel, in some order.
7. Two things we know about the child who brought the detective fiction novel.
    a. First, they sit next to Ellie.
    b. Second, they sit somewhere left of the student who brought "Playing the Odds."
8. The student who brought "The Last Day of Summer" sits somewhere left of the student who brought "For All the Happiness in the World."
9. The student who brought the graphic novel sits somewhere left of the one who brought "The Last Day of Summer."

Book Day

	Book Day	Desk					Book title					Genre				
		1	2	3	4	5	Finding Theo	For All the...	Playing the Odds	The Best of Jokes	The Last Day...	Action	Detective	Graphic novel	Horror	Sci-Fi
Name	Arnie															
	Cam															
	Ellie															
	Julie															
	Vince															
Genre	Action															
	Detective															
	Graphic novel															
	Horror															
	Sci-Fi															
Book title	Finding Theo															
	For All the...															
	Playing the Odds															
	The Best of Jokes															
	The Last Day...															

Name	Desk	Book title	Genre

All the Luck They Can Get

Exam week is here, and some of the students need all the luck they can get! These superstitious students have their exams on different days, for different subjects, and are each bringing their own lucky charm to—hopefully—help them beat the odds.

What is Jay's lucky charm, and what test does he need it for?

1. There is one day between the day the lucky rabbit's paw will be used and the day of the history test, in some order.
2. There is one day between the day of the math test and the day the lucky coin will be used, in some order.
3. Also, Gail will take her test sometime before the owner of the lucky coin has their exam.
4. Four things we know about the lucky die.
    a. First, it will be used either the day before or the day after the lucky jacket is used.
    b. Second, it will be used either the day before or the day after the biology test (although we don't know the charm used during this test).
    c. Third, it will be used sometime after Norm has taken his test.
    d. Fourth, it will be used sometime after Jay has taken his test.
5. Norm will take his test sometime before the owner of the lucky horseshoe.
6. Mabel will not take her test on Thursday, but has her exam sometime before the literature exam.

All the Luck They Can Get

	Weekday					Lucky charm					Test subject				
All the Luck They Can Get	Monday	Tuesday	Wednesday	Thursday	Friday	Coin	Die	Horseshoe	Jacket	Rabbit's paw	Biology	English	History	Literature	Math
Student Emma															
Gail															
Jay															
Mabel															
Norm															
Test subject Biology															
English															
History															
Literature															
Math															
Lucky charm Coin															
Die															
Horseshoe															
Jacket															
Rabbit's paw															

Student	Weekday	Lucky charm	Test subject

A Five-Step Password

The local bank has a maximum-security password for its vault, requiring five steps to unlock. On a special screen, shapes, colored squares, and words appear. For each step, you must input the correct shape, color, and word (the name of an animal), never repeating any shape, color, or word in subsequent steps.

What shape must you input alongside the word "bird," and what step must you input both items in?

1. The color purple and the star must be input in the same step, which occurs sometime before the step with a password requiring the color green.
2. We also know the step requiring color purple takes place sometime before the step with a password requiring the word "rabbit" (however, we do not know the color used with this word).
3. Additionally, there is one step between the step with the color purple and the step where you must input the word "cat", in some order.
4. The color yellow must be input in a step sometime earlier than the step where you must input the color blue, but in a step sometime later than the step where you must input the square and the step where you must input the moon.
5. The step where you must input the word "fish" takes place sometime after the step where you must input the moon.
6. Also, you must input the moon in a step sometime before you must input the star (however, we do not know what animal name comes with the star).
7. You must input the word "dog" in the step just before the step where you must input the circle.

A Five-Step Password

A Five-Step Password		Step					Color					Word				
		1	2	3	4	5	Blue	Green	Purple	Red	Yellow	Bird	Cat	Dog	Fish	Rabbit
Shape	Circle															
	Moon															
	Square															
	Star															
	Triangle															
Word	Bird															
	Cat															
	Dog															
	Fish															
	Rabbit															
Color	Blue															
	Green															
	Purple															
	Red															
	Yellow															

Shape	Step	Color	Word

A Week at the Colosseum

There's a lot of action at the Colosseum this week! Every day, Monday through Friday, a different gladiator will face a different animal using a different kind of weapon.

What animal will be fought with the bow and arrow, and on what day will this fight occur?

1. The gladiator fighting with a club will do so sometime before the day a gladiator will face a lion.
2. Also, the club will be used some day after the lance is used.
3. Two things we know about the bear fight.
    a. First, it will take place on an earlier day than the day the lance will be used.
    b. Second, there is one day between the day of this fight and the day the sword will be used, in some order.
4. Speaking of the sword, there is one day between the day it will be used and the day Rufus will fight, in some order.
5. There are two days between the day of the tiger fight and the day Marcellus will fight, in some order.
6. The tiger will also be fought some day earlier than the day a sling will be used.
7. There is one day between the day Cassius will fight and the day of the dog fight, in some order.
8. Theodorus will fight either the day after or the day before the lion fight.

A Week at the Colosseum

	A Week at the Colosseum	Weekday					Weapon					Animal				
		Monday	Tuesday	Wednesday	Thursday	Friday	Bow and arrows	Club	Lance	Sling	Sword	Bear	Dog	Lion	Tiger	Wolf
Gladiator	Caius															
	Cassius															
	Marcellus															
	Rufus															
	Theodorus															
Animal	Bear															
	Dog															
	Lion															
	Tiger															
	Wolf															
Weapon	Bow and arrows															
	Club															
	Lance															
	Sling															
	Sword															

Gladiator	Weekday	Weapon	Animal

Pizza With Everything!

At the pizza restaurant, Tables 1 through 5 are occupied by families having dinner. These tables are ordered in a row, in increasing numerical order, with Table 1 being the leftmost. Each has ordered a family pizza with two toppings, one meat and one non-meat, and a different soda pitcher for the family.

What's the second topping of the sausage pizza, and what soda was ordered with that pizza?

1. Three things we know about the family that ordered pizza with tomato.
   a. First, they're sitting somewhere to the right of the family that ordered the sausage pizza.
   b. Second, they are sitting next to the family that ordered cola.
   c. Third, there are two tables between the family that ordered the pizza with tomatoes and the family that ordered pizza with olives.
2. Three things we know about the family that ordered lemon-lime soda.
   a. First, they also ordered pizza with ham.
   b. Second, there is one table between the lemon-lime soda drinkers and the family that ordered ginger ale.
   c. Third, the family that ordered lemon-lime soda is sitting next to the family that ordered pizza with onions.
3. The family that ordered orange soda is sitting somewhere to the right of the family that ordered pizza with olives.
4. Two things we know about the family that ordered pizza with chicken.
   a. First, they're sitting somewhere to the left of the family that ordered ginger ale.
   b. Second, there is one table between the family who ordered chicken topping and the family that ordered bacon topping.
5. The family that ordered root beer is sitting somewhere to the right of the family that ordered pizza with mushrooms.

34

Pizza With Everything!

	Pizza With Everything!	Table					Soda					Non-meat				
		1	2	3	4	5	Cola	Ginger ale	Lemon-lime	Orange	Root beer	Mushrooms	Olives	Onions	Paprika	Tomato
Meat	Bacon															
	Chicken															
	Ham															
	Pepperoni															
	Sausage															
Non-meat	Mushrooms															
	Olives															
	Onions															
	Paprika															
	Tomato															
Soda	Cola															
	Ginger ale															
	Lemon-lime															
	Orange															
	Root beer															

Meat	Table	Soda	Non-meat

Avian Artistry

Tanya is an incredible artist, and birds are the primary inspiration for her art. These last months, April through August, she has produced five different pieces. Each depicts a different bird species, was created using a different technique, and was dedicated to a different person.

Who was the piece of art depicting a sparrow dedicated to, and what technique was used to create it?

1. Tanya photographed a bird sometime before creating the piece of art she dedicated to her brother, but sometime after creating both the raven and owl art pieces.
2. The work of art depicting a raven was made sometime after the art piece Tanya dedicated to her mom.
3. Three things we know about the piece of art depicting a woodpecker.
   a. First, it was in the cartoon style.
   b. Second, it was made either the month before or the month after Tanya dedicated a piece of art to her brother.
   c. Third, it was made sometime after Tanya dedicated a piece of artwork to her dad.
4. Two things we know about the oil painting.
   a. First, Tanya produced one work of art between the oil painting and the piece of art dedicated to her sister, in some order.
   b. Second, it was made sometime before the ink painting.
5. Tanya produced two works of art between the ink painting and the art piece depicting an eagle, in some order.

Avian Artistry

	Avian Artistry	Month					Technique					Dedicated to				
		April	May	June	July	August	Cartoon	Ink painting	Photograph	Oil painting	Sketch	Brother	Dad	Friend	Mom	Sister
Bird	Eagle															
	Owl															
	Raven															
	Sparrow															
	Woodpecker															
Dedicated to	Brother															
	Dad															
	Friend															
	Mom															
	Sister															
Technique	Cartoon															
	Ink painting															
	Photograph															
	Oil painting															
	Sketch															

Bird	Month	Technique	Dedicated to

The Five Temples

You must visit five temples on your quest to achieve human perfection. Each temple is protected by a different animal spirit and houses a different master who will teach you a different lesson. You must also visit the temples in a certain order.

When will you visit Master Raku, what will he teach you, and what spirit protects his temple?

1. Two things we know about Master Pahuk.
   a. First, you must visit his temple sometime after you've learned about wisdom.
   b. Second, there is one temple to visit between Master Pahuk's and Master Teshi's, in some order.
2. Speaking of Master Teshi, we know two things about him.
   a. First, he awaits you in the temple protected by the phoenix spirit.
   b. Second, you must visit Master Teshi's temple sometime before you learn about patience.
3. Two things we know about the temple protected by the eagle spirit.
   a. First, you must visit the eagle's temple sometime before you learn about wisdom.
   b. Second, there are two temples to visit between this one and the temple where you'll learn about bravery, in some order.
4. Two things we know about Master Josun.
   a. First, there is one temple visitation between the visit to his temple and the visit to the temple protected by the crane spirit, in some order.
   b. Second, there is one temple visitation between Master Josun's temple and the temple where you'll learn about strength, in some order (we still don't know what spirits protects this temple).
5. You must visit the temple where you'll learn generosity sometime after you have visited Master Gonji, but sometime before you visit the temple protected by the monkey spirit.
6. There is one temple visitation between the monkey spirit's temple and the crane spirit's temple, in some order.

38

The Five Temples

	The Five Temples	Visit number					Spirit					Lesson				
		1	2	3	4	5	Crane	Dragon	Eagle	Monkey	Phoenix	Bravery	Generosity	Patience	Strength	Wisdom
Master	Gonji															
	Josun															
	Pahuk															
	Raku															
	Teshi															
Lesson	Bravery															
	Generosity															
	Patience															
	Strength															
	Wisdom															
Spirit	Crane															
	Dragon															
	Eagle															
	Monkey															
	Phoenix															

Master	Visit number	Spirit	Lesson

Strange Locker Combinations

These lockers are very weird... To open them, you must input a code consisting of a number, a letter and a shape. There are five lockers in a row, starting with Locker 1 on the far left and increasing in numerical order.

What is the combination of Locker 5?

1. The locker with a code including a circle is next to the locker that opens with a code including a rectangle, in no specific order.
2. The locker with a code including the number 29 is somewhere to the left of the locker with a code including a rectangle, but somewhere to the right of the locker with a code including the number 37.
3. Four things we know about the locker with a code including the number 60.
   a. First, it is next to the one that opens with a code that includes a triangle.
   b. Second, it is somewhere to the left of the one with a code that includes the letter W.
   c. Third, it is somewhere to the right of the lockers that have codes including the letters Z and Q.
   d. Fourth, there is one locker between it and the locker with a code including the letter Y, in some order.
4. The locker with a code including the letter Y is somewhere to the left of the one with a code including the letter Q, and somewhere to the left of the locker with a code that includes a rhombus (though we're not sure what letter is included in the combination including a rhombus.)
5. The locker with a code including the number 82 is somewhere to the right of the locker with a code including the number 94.

Strange Locker Combinations

	Strange Locker Combinations	Locker					Shape					Letter				
		1	2	3	4	5	Circle	Rectangle	Rhombus	Square	Triangle	Q	W	X	Y	Z
Number	29															
	37															
	60															
	82															
	94															
Letter	Q															
	W															
	X															
	Y															
	Z															
Shape	Circle															
	Rectangle															
	Rhombus															
	Square															
	Triangle															

Number	Locker	Shape	Letter

Sacrifices to the Gods

Five people have made offerings to the Gods this week. Each of them has made a different offering at a different temple, and prays that the animal they offered will prove worthy of a blessing.

What animal did Zoe sacrifice, on which day, and to which god?

1. The horse was offered to the gods on a later day than Athena's temple was visited.
2. Three things we know about the sacrificed bull.
   a. First, it was offered to the gods either the day before or the day after the goat.
   b. Second, it was offered to the gods at some point before the temple of Apollo was visited.
   c. Third, there are two days between the one on which the bull was sacrificed and the day Theodore visited a temple, in some order.
3. Two things we know about the temple of Zeus.
   a. First, someone visited it either the day before or the day after the goose was sacrificed.
   b. Second, it was visited at some point before the sheep was sacrificed, and also sometime before Phoebe visited one of the temples (though we don't know what animal Phoebe sacrificed).
4. Someone visited the temple of Poseidon at some point before the temple of Apollo was visited, but sometime after the sheep was sacrificed.
5. There is one day between the day the temple of Artemis was visited and the day Constantine visited another of the temples, in some order.
6. Daphne visited a temple either the day before or the day after the goat was sacrificed.

Sacrifices to the Gods

		Weekday					Temple					Offering				
Sacrifices to the Gods		Monday	Tuesday	Wednesday	Thursday	Friday	Apollo	Artemis	Athena	Poseidon	Zeus	Bull	Goat	Goose	Horse	Sheep
Name	Constantine															
	Daphne															
	Phoebe															
	Theodore															
	Zoe															
Offering	Bull															
	Goat															
	Goose															
	Horse															
	Sheep															
Temple	Apollo															
	Artemis															
	Athena															
	Poseidon															
	Zeus															

Name	Weekday	Temple	Offering

Driver App Evals

The new driving app prides itself on quality drivers! Here are five of them, with 4.6, 4.7, 4.8, 4.9 and 5.0 stars respectively. They drive different brands of car, each with a different license plate.

What's Lydia's car's license plate number, and what is her rating as a driver?

1. Ronnie drives a Peugeot and has a higher rating than the Chevrolet driver.
2. Both Betty and Darius have higher ratings than the Subaru driver.
3. The Chevrolet driver has a lower rating than Casper.
4. There is a 0.2 star difference between the Mitsubishi driver and the driver of the car with the license plate TKI327.
5. Also, the driver of the car with license plate TKI327 has a higher rating than the driver of the car with the license plate QSD580, but a lower rating than the Citroën driver.
6. The Citroën driver has a higher score than the driver of the car with license plate JKR143.
7. Also, there is a 0.2 star difference between the driver of the car with license plate JKR143 and Darius.
8. The driver of the car with license plate PJZ458 has a lower rating than the Chevrolet driver.

Driver App Evals

	Driver App Evals	Rating					Plate					Car				
		4.6	4.7	4.8	4.9	5.0	HYU792	JKR143	PJZ458	QSD580	TKI327	Chevrolet	Citröen	Mitsubishi	Peugeot	Subaru
Name	Betty															
	Casper															
	Darius															
	Lydia															
	Ronnie															
Car	Chevrolet															
	Citröen															
	Mitsubishi															
	Peugeot															
	Subaru															
Plate	HYU792															
	JKR143															
	PJZ458															
	QSD580															
	TKI327															

Name	Rating	Plate	Car

The Rose Contest

The annual rose contest was a success! Many beautiful flowers were appreciated, and everybody had a great time. Each of the first five places was taken by a rose of a different type and different color.

What color was the Grandiflora rose, and what place did it take?

1. One woman placed between Mrs. Morris and Mrs. Dhalmar, in some order.
2. By the way, the white rose beat the rose presented by Mrs. Dhalmar.
3. Mrs. Potter presented the yellow rose.
4. Two things we know about the Damask rose.
   a. First, it took either the place just below or just above the pink rose.
   b. Second, it did not take third place.
5. The pink rose placed somewhere below the Hybrid Musk rose.
6. The white rose placed just above or just below the Hybrid Tea rose.
7. The Floribunda rose is orange, and placed somewhere below Mrs. Thompson's rose in the contest.
8. Mrs. Thompson also placed somewhere below Mrs. Dhalmar, but somewhere above Mrs. Griffin.

The Rose Contest

The Rose Contest		Place					Species					Color				
		1	2	3	4	5	Damask	Grandiflora	Floribunda	Hybrid Musk	Hybrid Tea	Orange	Pink	Red	White	Yellow
Owner	Mrs. Dhalmar															
	Mrs. Griffin															
	Mrs. Morris															
	Mrs. Potter															
	Mrs. Thompson															
Color	Orange															
	Pink															
	Red															
	White															
	Yellow															
Species	Damask															
	Grandiflora															
	Floribunda															
	Hybrid Musk															
	Hybrid Tea															

Owner	Place	Species	Color

Slot Machine Madness

In this casino, Slot Machines 1 through 5 are located in ascending numerical order, side by side along a corridor, with Machine 1 being the leftmost. Each machine has a different shape, fruit, and animal that can appear in the slots. Any of the images repeated three times in a row guarantees a payout!

What are the three symbols on Slot Machine 2?

1. The slot machine with bird designs is somewhere left of the machine with stars, but somewhere right of the one with mice designs.
2. The slot machine with bird designs is also somewhere left of the machine featuring lemons, though we don't know the other two symbols on that machine.
3. Two things we know about the slot machine with moon designs.
   a. First, there is one slot machine between the machine that features moon designs and the machine that features apple designs, with moons being to the left of apples.
   b. Second, the moon design slot machine is somewhere to the left of the machine that features cherry designs.
4. The slot machine with sun designs is somewhere left of the machine that has triangle designs.
5. The slot machine with grape designs is somewhere right of the machine that has cat designs, but somewhere left of the one that has heart designs.
6. Three things we know about the slot machine with fish designs.
   a. First, there is one slot machine between it and the machine with triangle designs, in some order.
   b. Second, it is next to the machine that has cherry designs.
   c. Third, it is next to the machine that features star designs (but we do not know the other symbols on that machine).

Slot Machine Madness

Slot Machine Madness		Number					Fruit					Animal				
		1	2	3	4	5	Apples	Cherries	Grapes	Lemons	Oranges	Birds	Cats	Dogs	Fish	Mice
Shape	Hearts															
	Moons															
	Stars															
	Suns															
	Triangles															
Animal	Birds															
	Cats															
	Dogs															
	Fish															
	Mice															
Fruit	Apples															
	Cherries															
	Grapes															
	Lemons															
	Oranges															

Shape	Number	Fruit	Animal

Cellblock Countdown

Cells 1 through 5 of the city's maximum-security prison for men house five hardened criminals. The cells are along a corridor, with Cell 1 on the far left and Cell 5 on the far right. Each of the men has a different eye and hair color, and also a scar on a different part of the body.

Where does the red-haired inmate have a tattoo, and what cell is he jailed in?

1. The inmate with amber eyes is in a cell somewhere to the left of the inmate with blue eyes.
2. Two things we know about the inmate with a scar on his leg.
   a. First, he is in a cell adjacent to the inmate with amber eyes.
   b. Second, his cell is somewhere to the right of the inmate with green eyes, but somewhere to the left of the inmate with gray eyes.
3. Three things we know about the inmate with gray hair.
   a. First, he's in a cell somewhere to the right of the inmate with black hair.
   b. Second, there is one cell between the cell where the gray-haired inmate is and the cell holding the inmate with a scar on his neck.
   c. Third, there are two cells between the one where he is and the cell containing the inmate with gray eyes.
4. The inmate with a scar on his neck and the one with a scar on his face are in adjacent cells.
5. Two things we know about the red-haired inmate.
   a. First, he is in a cell somewhere to the right of the inmate with a scar on his hand.
   b. Second, he's in a cell adjacent to that of the gray-eyed inmate.
6. The inmate with a scar on his hand has blond hair.

50

Cellblock Countdown

Cellblock Countdown	Prison cell					Scar location					Eye color				
	1	2	3	4	5	Arm	Face	Hand	Leg	Neck	Amber	Blue	Gray	Green	Hazel
Hair color Black															
Blond															
Brown															
Gray															
Red															
Eye color Amber															
Blue															
Gray															
Green															
Hazel															
Scar location Arm															
Face															
Hand															
Leg															
Neck															

Hair color	Prison cell	Scar location	Eye color

Furniture Purchases

The local furniture store creates custom furniture to fit all kinds of homes. Today, five items were sold to five different families. Each item was a different type of furniture and had its own model name. Items come in the following sizes, from smallest to largest: XS, S, M, L, and XL.

What family bought the Alexia model and what type of furniture was it?

1. There is one item that fits between the McQueen family's purchase and the Halley model in terms of size.
2. There are two items that fit between the Rubicon model and the Brenton family's purchase in terms of size.
3. The Abbott family's purchase is some size smaller than the armchair.
4. There is one item that fits between the bed and the Roman model in terms of size.
5. Three things we know about the chair.
   a. First, it is some size smaller than the Jackson family's purchase.
   b. Second, it is some size larger than the Rubicon model.
   c. Third, it's either one size smaller or one size larger than the bed.
6. The Jasper model is some size larger than the Halley model.
7. At the same time, the sofa is some size larger than the Jasper model.
8. The bed is one size smaller or one size larger than the armchair.
9. There is one model that first between the piece of furniture the Jackson family bought and the sofa in terms of size.
10. The McQueen family's purchase is smaller than the Jasper model.

Furniture Purchases

	Furniture Purchases	Size Category					Model					Type				
		XS	S	M	L	XL	Alexia	Halley	Jasper	Roman	Rubicon	Armchair	Bed	Chair	Sofa	Table
Family	Abbott															
	Brenton															
	Clarke															
	Jackson															
	McQueen															
Type	Armchair															
	Bed															
	Chair															
	Sofa															
	Table															
Model	Alexia															
	Halley															
	Jasper															
	Roman															
	Rubicon															

Family	Size	Model	Type

Consulting the Gods

Ancient Romans often consulted the gods before making big decisions, and the five in this puzzle are no different! Each of these Roman priests sought advice on a different day, for a different reason, and used a different method of interpretation.

What did Julius consult the gods about, and what method did he use to interpret the gods' advice?

1. Five things we know about Attilius.
   a. First, he consulted the gods either the day before or the day after the priest that consulted about a war.
   b. Second, he consulted sometime later than the priest who consulted about sailing.
   c. Third, he consulted either the day before or the day after Octavius (though we still don't know what either of them consulted about).
   d. Fourth, there was one day between the day he consulted the gods and the day a priest consulted about business, in some order.
   e. Fifth, he consulted the gods sometime earlier than the priest pondering politics.
2. The priest consulted about politics sometime earlier than the priest who analyzed animal entrails, and also either the day after or the day before the priest who observed weather patterns.
3. The priest who consulted an oracle did so sometime later than the priest who observed the weather.
4. Three things we know about Marcus.
   a. First, there was one day between the day the oracle was consulted and the day Marcus consulted the gods, in some order.
   b. Second, Marcus consulted the gods on a later day than the priest who observed birds in flight.
   c. Third, Marcus consulted the gods sometime earlier than Quintus.

Consulting the Gods

Consulting the Gods		Weekday					Topic					Method				
		Monday	Tuesday	Wednesday	Thursday	Friday	Business	Politics	Sailing	War	Wedding	Animal entrails	Bird flight	Chicken feed	Oracle	Weather patterns
Name	Attilius															
	Julius															
	Octavius															
	Marcus															
	Quintus															
Method	Animal entrails															
	Bird flight															
	Chicken feed															
	Oracle															
	Weather patterns															
Topic	Business															
	Politics															
	Sailing															
	War															
	Wedding															

Name	Weekday	Topic	Method

The Sticker Collection

A new sticker collection album contains stickers of various famous people of history. Lysander is swapping all of his duplicate stickers with his friends, to try to collect them all. He swapped stickers with a different friend each day, in exchange for a sticker with a different chrome color and a different famous person from history.

What color sticker did Lysander get from Claudius, what famous person was on that sticker, and what day did Lysander and Claudius swap stickers?

1. Three things we know about the sticker of Julius Caesar.
    a. First, Lysander got it before he got the gold sticker.
    b. Second, he got it either the day before or the day after he got the sticker with Joan of Arc on it.
    c. Third, he got two other stickers between getting the Julius Caesar sticker and getting the sticker with Napoleon on it, in some order.
2. Speaking of Napoleon, Lysander got his sticker before getting the sticker of Henry VIII.
3. Two things we know about the red sticker.
    a. First, Lysander got one other sticker between the red sticker and the sticker he got from Claudius, in some order.
    b. Second, he got the red sticker sometime before the pink sticker.
4. The swap with Titus took place at some point before Lysander got the white sticker.
5. Two things we know about the orange sticker.
    a. First, Lysander got the orange sticker and the white sticker on consecutive days, in some order.
    b. Second, there are two days between the day he got the orange sticker and the day he got the sticker with Empress Elizabeth on it, in some order.
6. There is one day between the day Lysander swapped stickers with Hamlet and the day he swapped stickers with Cassandra. The swap with Cassandra came first.

The Sticker Collection

	Weekday					Person					Color				
The Sticker Collection	Monday	Tuesday	Wednesday	Thursday	Friday	Empress Elizabeth	Henry VIII	Joan of Arc	Julius Caesar	Napoleon	Gold	Orange	Pink	Red	White
Friend															
Cassandra															
Claudius															
Hamlet															
Ophelia															
Titus															
Color															
Gold															
Orange															
Pink															
Red															
White															
Person															
Empress Elizabeth															
Henry VIII															
Joan of Arc															
Julius Caesar															
Napoleon															

Friend	Weekday	Person	Color

Boggling Backpacks

At school, the first row contains five desks, starting with Desk 1 at the far left and increasing in numerical order. As it happens, all students sitting in those desks have the coolest backpacks in the class! Each backpack is a different color and features a different design.

What color is Gary's backpack and what's its cool design?

1. The orange backpack belongs to someone sitting somewhere left of the brown backpack's owner.
2. Similarly, the owner of the brown backpack sits somewhere left of Kelly.
3. Vanessa sits next to the owner of the blue backpack.
4. Three things we know about the purple backpack.
    a. First, its owner sits somewhere to the left of the owner of the dog backpack, but somewhere to the right of the owner of the dinosaur backpack.
    b. Second, the purple backpack's owner sits to the immediate left of the owner of the tiger backpack.
    c. Third, the owner of the purple backpack sits next to the owner of the yellow backpack.
5. The dinosaur backpack belongs to someone sitting somewhere left of Eddie.
6. Speaking of Eddie, he sits somewhere left of Brenda.
7. And speaking of Brenda, she sits next to the owner of the yellow backpack.
8. The lion backpack belongs to a kid sitting next to Kelly.

Boggling Backpacks

	Boggling Backpacks	Seat					Color					Design				
		1	2	3	4	5	Blue	Brown	Orange	Purple	Yellow	Bear	Dinosaur	Dog	Lion	Tiger
Child	Brenda															
	Eddie															
	Gary															
	Kelly															
	Vanessa															
Design	Bear															
	Dinosaur															
	Dog															
	Lion															
	Tiger															
Color	Blue															
	Brown															
	Orange															
	Purple															
	Yellow															

Child	Seat	Color	Design

Summer Fun

This family sure knows how to enjoy summer! Each day of the week, one of the members of the household gets to choose an indoor game and an outdoor sport for the whole family to enjoy.

When did the family go swimming, what else did they do that day, and who picked the day's activities?

1. Three things we know about the grandmother.
   a. First, she got to plan the day either the day before or the day after the family played liar's dice.
   b. Second, she got to plan the day at some point before the daughter was the planner.
   c. Third, the grandmother got to pick the activities at some point after the family played Scrabble.
2. Two things we know about the day the family went biking.
   a. First, it was sometime before the day they went hiking.
   b. Second, it was sometime later than the day they played football.
3. Two things we know about the dad.
   a. First, there was one day between the day he planned and the day the family played basketball, in some order.
   b. Second, he planned activities either the day before or the day after the family played blackjack.
4. The family played dominoes at some point after the day they played blackjack, and also at some point after the day the son planned (though we don't know what plans he made).
5. There are two days between the day the family played Scrabble and the day the mom planned, in some order.
6. The family played football the day before the daughter decided the activities.

Summer Fun

Summer Fun	Weekday					Game					Sport				
	Monday	Tuesday	Wednesday	Thursday	Friday	Blackjack	Dominoes	Liar's Dice	Monopoly	Scrabble	Basketball	Biking	Football	Hiking	Swimming
Planner Dad															
Daughter															
Grandmother															
Mom															
Son															
Sport Basketball															
Biking															
Football															
Hiking															
Swimming															
Game Blackjack															
Dominoes															
Liar's Dice															
Monopoly															
Scrabble															

Planner	Weekday	Game	Sport

The Tattoo Artist

The local tattoo artist has five different clients scheduled, each coming on a different day of the week to get a different design tattooed on a different body part.

What tattoo does Chuck want, and on what day is he scheduled to get it?

1. There are two days between the day the arm will be tattooed and the day the shoulder will be tattooed, in some order.
2. Also, the skull will be tattooed either the day before or the day after the shoulder will be tattooed.
3. There is one day between the day of Dave's appointment and the day a chest will be tattooed, in some order.
4. Similarly, there are two days between the day the leg will be tattooed and the day the chest will be tattooed, in some order.
5. Speaking of the chest tattoo, it's scheduled for either the day before or the day after Garrett's appointment.
6. The arm will be tattooed either the day before or the day after the day of Ava's appointment.
7. The client who wants the rose tattoo will be coming some day after Ava.
8. There is one day between the day a yin-yang symbol will be tattooed and the day the rose will be tattooed, in some order.
9. Sue will be coming on a later day than the client who wants a skull tattoo.
10. Also, the skull tattoo is scheduled for either the day before or the day after the leg tattoo.
11. The person who wants the name of their child tattooed is coming sometime later than the one who wants a tattoo of a dragon.

The Tattoo Artist

The Tattoo Artist		Weekday					Design					Location				
		Monday	Tuesday	Wednesday	Thursday	Friday	Dragon	Name	Rose	Skull	Yin-yang	Arm	Chest	Leg	Neck	Shoulder
Client	Ava															
	Chuck															
	Dave															
	Garrett															
	Sue															
Location	Arm															
	Chest															
	Leg															
	Neck															
	Shoulder															
Design	Dragon															
	Name															
	Rose															
	Skull															
	Yin-yang															

Client	Weekday	Design	Location

The Journey Through Town

You're on your way to meet your friends, but the instructions to get to the meetup place are quite complex. In the step-by-step instructions your friends wrote, you follow different streets until you find certain shops, owned by certain shopkeeps. You'll arrive at your destination only after you've followed all five steps.

What establishment does Theo own, and what step of your journey do you finish when you find his shop?

1. Your journey along Sweeney Street must take place sometime earlier than the step that leads you to the bakery.
2. It also takes place either just before or just after you travel along Acorn Street.
3. By the way, you must walk along Gladwell Street sometime after you travel down Acorn Street.
4. Three things we know about the step that leads you to the butcher's shop.
   a. One, it takes place sometime before the step that leads you to Claudia's establishment.
   b. Second, there is one step between the step that leads to the butcher shop and the step that leads to Angela's establishment, in some order.
   c. Third, you'll walk by the butcher shop sometime after you travel along Sweeney Street.
5. Two things we know about Millboy Street.
   a. First, it is not the last road of the journey.
   b. Second, you'll walk along it sometime later than you walk by the bakery.
6. The step that takes you past the jeweler's is either just before or just after the step that takes you along Acorn Street.
7. There are two steps between the step that passes by Adam's establishment and the step that passes by Brian's establishment, with Brian's shop appearing later in the journey than Adam's.
8. Similarly, there is one step between the step that takes you past Adam's establishment and the step that takes you along Acorn Street, in some order.
9. You'll pass by Claudia's establishment sometime later than you travel past the inn.

The Journey Through Town

The Journey Through Town		Step					Shop					Shopkeep				
		1	2	3	4	5	Bakery	Barber	Butcher	Inn	Jeweler	Adam	Angela	Brian	Claudia	Theo
Street	Acorn															
	Frazer															
	Gladwell															
	Millboy															
	Sweeney															
Shopkeep	Adam															
	Angela															
	Brian															
	Claudia															
	Theo															
Shop	Bakery															
	Barber															
	Butcher															
	Inn															
	Jeweler															

Street	Step	Shop	Shopkeep

A Rainy Week

It's been raining all week, and these five friends have taken turns hosting indoor events where they hang out and enjoy indoor activities. The host is also responsible, of course, for providing something for their guests to eat and drink.

When did Sawyer host his friends, and what food and drink did he offer them for their evening snack?

1. Four things we know about Kinsley.
   a. First, she hosted her friends sometime after the group snacked on cookies.
   b. Second, there is one day between the day Kinsley hosted and the day the group drank hot cocoa, in some order.
   c. Third, Kinsley hosted sometime after spiced wine was served.
   d. Fourth, Kinsley hosted either the day before or the day after they had brownies.
2. Two things we know about the day coffee was served.
   a. First, there was one day between coffee day and the day waffles were offered, in some order.
   b. Second, the guests drank coffee at some point after drinking hot cocoa.
3. Speaking of hot cocoa, it was offered to the guests sometime after the hot milk.
4. Rowan did not host his friends on Thursday.
5. Two things we know about Leilani.
   a. First, she hosted her friends at some point after they all drank tea.
   b. Second, she served everyone orange cake.
6. The waffles were offered sometime after Parker hosted.

A Rainy Week

	Weekday					Drink					Food				
A Rainy Week	Monday	Tuesday	Wednesday	Thursday	Friday	Coffee	Hot cocoa	Hot milk	Spiced wine	Tea	Apple pie	Brownies	Cookies	Orange cake	Waffles
Host Kinsley															
Leilani															
Parker															
Rowan															
Sawyer															
Food Apple pie															
Brownies															
Cookies															
Orange cake															
Waffles															
Drink Coffee															
Hot cocoa															
Hot milk															
Spiced wine															
Tea															

Host	Weekday	Drink	Food

Card Tricks

Fiona loves learning card tricks, and has delighted the family with a new trick each day this week. The card was affected differently each time, her family selected different cards each day, and Fiona learned each trick from a different source.

On what day was the ace of spades selected, and where did Fiona learn the trick she performed that day?

1. Fiona performed the trick she learned from a friend sometime earlier than she made a card appear in her spectator's pocket, but sometime later than she performed the trick from the magic book.
2. Four things we know about the three of spades.
    a. First, it was selected either the day before or the day after the seven of hearts was chosen.
    b. Second, it was part of a trick performed sometime earlier in the week than the trick Fiona learned from her dad.
    c. Third, there is one day between the day the three of spades was chosen and the day Fiona performed the trick she learned from YouTube, in some order.
    d. Fourth, the three of spades was chosen either the day before or the day after Fiona made the chosen card disappear from the deck (and we still don't know which card that was).
3. The nine of diamonds was selected sometime later in the week than the day Fiona performed the trick she learned from her dad, and also sometime later in the week than she performed the trick she learned from YouTube.
4. Two things we know about the trick in which the selected card appears on top of the deck.
    a. First, there are two days between the day Fiona made the card appear on top of the deck and the day the king of clubs was selected, in some order.
    b. Second, she made the card appear on top of the deck the day before she made the card show up face down in the deck.
5. The king of clubs was selected sometime after Fiona performed the trick she learned from YouTube.

Card Tricks

Card Tricks	Weekday					Effect					Source				
	Monday	Tuesday	Wednesday	Thursday	Friday	Appears in hand	Appears in pocket	Disappears from deck	Face down in deck	On top of deck	Book	Dad	Friend	Magic lesson	YouTube
Card Chosen Ace of Spades															
Three of Spades															
Seven of Hearts															
Nine of Diamonds															
King of Clubs															
Source Book															
Dad															
Friend															
Magic lesson															
YouTube															
Effect Appears in hand															
Appears in pocket															
Disappears from deck															
Face down in deck															
On top of deck															

Card chosen	Weekday	Effect	Source

Start Your Engines

A great car race is about to take place. All participants are talented and experienced drivers who each won the national cup in a different country. Each of them is racing a different colored car in their designated lane, numbered 1-5 with Lane 1 on the far left of the track. It's sure to be an action-packed day!

What color is Bernard's car and what lane is he driving in?

1. Three things we know about Carl.
   a. First, he's in a lane somewhere to the left of the man who won the USA's national cup.
   b. Second, he's in the lane directly right of the driver of the green car.
   c. Third, his line is either directly to the left or directly to the right of the man who won the Brazilian National Cup.
2. The man who won Denmark's national cup is in a lane somewhere to the right of Dwayne.
3. Speaking of Dwayne, he and Sheldon will race in adjacent lanes.
4. There was exactly one lane between the white and blue car's lanes, in some order.
5. Speaking of the driver of the blue car, he will race in a lane somewhere right of the winner of the USA National Cup.
6. The man who won the Brazilian National Cup will race in a lane somewhere to the left of Bernard.
7. Bernard, meanwhile, will race in a lane adjacent to the driver of the purple car.
8. The man who won the Canadian National Cup is in a lane somewhere right of the USA National Cup winner.
9. The red car's driver is in a lane adjacent to the Denmark National Cup winner.

70

Start Your Engines

Start Your Engines		Lane					Car					National				
		1	2	3	4	5	Blue	Green	Purple	Red	White	Brazil	Canada	Denmark	Germany	USA
Driver	Bernard															
	Carl															
	Dwayne															
	Sheldon															
	Toby															
National	Brazil															
	Canada															
	Denmark															
	Germany															
	USA															
Car	Blue															
	Green															
	Purple															
	Red															
	White															

Driver	Lane	Car	National

The Aquarium Dolphins Riddle

It's a glorious summer week, and the city aquarium is offering daily dolphin shows! Each day a different dolphin performs a different trick, and their trainers use a different type of food to entice them, since each dolphin has a different favorite food.

What is Shamu's trick, and what day of the week will he be performing?

1. The dolphin who jumps through hoops will perform sometime later in the week than the dolphins who love salmon and mussels (who will perform in some order), but sometime before Captain.
2. Three things we know about Rocky.
   a. First, he will perform sometime later than the dolphin who will give a high five.
   b. Second, he loves shrimp.
   c. Third, he'll perform either the day before or the day after the dolphin who loves squid.
3. The dolphin that will lift his trainer will perform sometime earlier in the week than the dolphin who loves shrimp.
4. Also, the trick of lifting the trainer and the trick of bouncing a ball will be performed on consecutive days, in some order.
5. Flipper will perform sometime before the dolphin that loves mussels, and also sometime before the one who will give a high five (although we don't know what the high-fiving dolphin loves to eat).
6. The dolphin that gives a high five will perform sometime before Wally performs.

The Aquarium Dolphins Riddle

	Weekday					Trick					Treat				
The Aquarium Dolphins Riddle	Monday	Tuesday	Wednesday	Thursday	Friday	Bounce ball	High five	Hoop jump	Lift trainer	Tail wave	Mussels	Salmon	Shrimp	Squid	Tuna
Name Captain															
Flipper															
Rocky															
Shamu															
Wally															
Treat Mussels															
Salmon															
Shrimp															
Squid															
Tuna															
Trick Bounce ball															
High five															
Hoop jump															
Lift trainer															
Tail wave															

Name	Weekday	Trick	Treat

The Birthday Celebrations Riddle

Luke has always enjoyed celebrating his birthday with a nice lunch, a special activity, and a present for himself. From ages 11 to 15, the lunch, activity, and present were different each time.

What present did Luke choose for his 15th birthday, and what lunch and activity did he celebrate that birthday with?

1. Three things we know about the jigsaw puzzle.
   a. First, there are two years between the year this present was given and the year Luke had tacos for lunch, in some order.
   b. Second, Luke got this gift either the year before or the year after he had pizza for lunch.
   c. Third, he got the jigsaw puzzle some year prior to the year he played soccer, but some year after he got a book.
2. There is one year between the year Luke went to the pool and the year he got the soccer ball, in some order.
3. He received the art set on a birthday sometime before getting the soccer ball.
4. Luke did not have a burger for lunch on his 12th birthday.
5. He did enjoy a burger on an earlier birthday than the year he visited the museum.
6. The rotisserie chicken, meanwhile, was enjoyed on a later year than the museum visit.
7. Luke got his toy robot either the birthday before or after he went to the theater performance.
8. There was one year between the year Luke had sushi for lunch and the year he got an art set, in some order.

The Birthday Celebrations Riddle

		Age					Activity					Present				
The Birthday Celebrations Riddle		11	12	13	14	15	Movie	Museum	Pool party	Soccer	Theater	Art set	Soccer ball	Book	Jigsaw puzzle	Toy robot
Lunch	Burgers															
	Pizza															
	Chicken															
	Sushi															
	Tacos															
Shape	Art set															
	Soccer ball															
	Book															
	Jigsaw puzzle															
	Toy robot															
Activity	Movie															
	Museum															
	Pool party															
	Soccer															
	Theater															

Lunch	Age	Activity	Present

Dragons and Gemstones

You're in front of five mysterious caves. Let's number them Caves 1 through 5, in order, with Cave 1 being the leftmost cave. In each cave there is a dragon of a different color, guarding a different gemstone.

What's the name and color of the dragon guarding the sapphire?

1. Two things we know about the red dragon.
    a. First, there are two caves between the red dragon's cave and the cave containing the emerald.
    b. Second, it lives somewhere to the right of the blue dragon.
2. Three things we know about Tipur.
    a. First, he lives in a cave somewhere to the left of the blue dragon.
    b. Second, he lives beside Fryn.
    c. Third, he lives in the cave immediately to the right of the dragon guarding the ruby.
3. Khamu lives somewhere to the left of the dragon guarding the amethyst.
4. The green dragon lives somewhere to the left of the black one.
5. Two things we know about the dragon guarding the diamond.
    a. First, it lives next door to Garwyn.
    b. Second, it lives next door to the white dragon (although we don't know the white dragon's name).
6. Nelbon lives somewhere to the left of Fryn.

Dragons and Gemstones

	Dragons and Gemstones	Cave					Gemstone					Color				
		1	2	3	4	5	Amethyst	Diamond	Emerald	Ruby	Sapphire	Black	Blue	Green	Red	White
Name	Fryn															
	Garwyn															
	Khamu															
	Nelbon															
	Tipur															
Color	Black															
	Blue															
	Green															
	Red															
	White															
Gemstone	Amethyst															
	Diamond															
	Emerald															
	Ruby															
	Sapphire															

Name	Cave	Gemstone	Color

Five Chess Sets

Who said chess sets always have to look the same? Chester Mason loves making chess sets, and crafts a new one every month, adding a unique twist to each set. like using different colors for the light and dark pieces and adding new pieces for players to have fun with. From June through October, Chester made sets in different colors, creating a different fantasy theme for every set's pieces.

What colors are the fairy set's dark and light pieces, and what month was that chess set made?

1. Three things we know about the chess set with blue pieces.
   a. First, it was made sometime after the wizard set.
   b. Second, it was made either the month before or the month after the month whose set featured brown pieces.
   c. Third, two other chess sets were made between the blue piece set and the beige piece set.
2. Two chess sets were made between the fairy set and the set with pink pieces, in some order.
3. The chess set with the orange pieces was made sometime after the set with the indigo pieces.
4. The wolf set was made either the month before or the month after the set with the red pieces was created.
5. Two things we know about the chess set with indigo pieces.
   a. First, it was made at some point before the set with the green pieces, and either the month before or the month after the wolf set (though we don't know what color the wolf set was painted).
   b. Second, two chess sets were made between the set with indigo pieces and the set with pink pieces, in some order.
6. The elf set was made sometime before the wolf set.

Five Chess Sets

	Five Chess Sets	Month					Light pieces					Dark pieces				
		June	July	August	September	October	Beige	Orange	Pink	Red	Yellow	Blue	Brown	Green	Indigo	Violet
Set style	Dragon															
	Elf															
	Fairy															
	Wolf															
	Wizard															
Dark pieces	Blue															
	Brown															
	Green															
	Indigo															
	Violet															
Light pieces	Beige															
	Orange															
	Pink															
	Red															
	Yellow															

Set Style	Month	Light Pieces	Dark Pieces

Cozy Caves

Five cavemen inhabit five caves, labeled in ascending numerical order starting with Cave 1 on the far left. Each of the caves is a veritable home, complete with a rug, a nice rock painting, and even some food in storage. The cavemen refuse to copy their neighbors, so each rug is made of a different type of animal skin, every rock painting depicts something different, and they all store a different type of food.

What rug is in the cave with the painting of a hunt, and what food is being stored inside?

1. The cave storing meat is next to the cave storing nuts.
2. Two things we know about the cave with the painting of the hunt.
    a. First, there is one cave between it and the one with nuts stored inside.
    b. Second, the cave with the painting of the hunt is the only cave between the one with the painting of the lake and the one with the painting of a night sky, in some order.
3. We know two things about the cave with the night sky painting.
    a. First, it is somewhere to the right of the caves with mammoth and hare skin rugs.
    b. Second, there are two caves between the cave with the painting of the night sky and the cave with roots stored within.
4. The cave with the horse skin rug is next door to the cave with a painting of the caveman's family.
5. It's also somewhere to the left of the cave with fruit stored in it (though we don't know what painting is in the fruit-stocked cave).
6. There is one cave between the cave with the mammoth skin rug and the cave with the cow skin rug.

Cozy Caves

Cozy Caves		Cave					Rock painting					Food				
		1	2	3	4	5	Family	Hunt	Lake	Night sky	Ritual	Fruit	Meat	Nuts	Roots	Vegetables
Rug skin	Cow															
	Deer															
	Hare															
	Horse															
	Mammoth															
Food	Fruit															
	Meat															
	Nuts															
	Roots															
	Vegetables															
Rock painting	Family															
	Hunt															
	Lake															
	Night sky															
	Ritual															

Rug skin	Cave	Rock painting	Food

The Five Newborns

Five beautiful babies were born today, and they're resting side by side in the hospital nursery. Their cradles are labeled numerically, with Cradle 1 being the leftmost. Each baby has a different name, a different hair color, and is wearing a different-patterned onesie.

What color hair does baby Caspian have, what onesie is he wearing, and what number cradle is he using?

1. There are two babies between the red-haired baby and blonde baby.
2. Two things we know about the brown-haired baby.
    a. First, it is directly to the left of the baby with the animals on their onesie.
    b. Second, the brown-haired baby's cradle is next to baby Maeve's cradle, in some order (but we know nothing about the onesie baby Maeve is wearing).
3. Two things we know about the baby wearing a onesie with rockets.
    a. First, there are two babies between it and the baby wearing a onesie with trains on it.
    b. Second, the baby with rockets on their onesie is resting somewhere to the left of the black-haired baby.
4. The baby wearing a onesie with teddy bears is the only one between baby Silas and baby Arlo, in some order.
5. Baby Isla is somewhere to the right of the baby wearing a onesie with clowns on it, but somewhere to the left of the blonde baby.
6. She is also somewhere to the left of baby Silas and baby Maeve (whose hair colors we don't know).

The Five Newborns

The Five Newborns	Cradle					Hair color					Onesie pattern				
	1	2	3	4	5	Bald	Blonde	Brown	Black	Red	Animals	Clowns	Rockets	Teddy bears	Trains
Name Arlo															
Caspian															
Isla															
Maeve															
Silas															
Onesie pattern Animals															
Clowns															
Rockets															
Teddy bears															
Trains															
Hair color Bald															
Blonde															
Brown															
Black															
Red															

Name	Cradle	Hair color	Onesie pattern

The Docks at Pirate Bay

At Pirate Bay, the docks are filled with the ships of five of the best pirate captains to sail the seven seas. Dock 1 is on the far left, and the docks increase numerically, proceeding to the right of the pier. Each ship has its own unique name, each of the captains have a distinctive physical mark, and each ship carries a different type of loot.

What treasure is being carried by the gold-toothed pirate captain, and what's the name of his ship?

1. The ship belonging to the captain with a nose earring is moored in Dock 2.
2. The ship belonging to the captain with a scar on his face is docked next to the Wave Crusher.
3. By the way, the Wave Crusher is docked somewhere to the right of the Sea Demon.
4. The Night Watcher is docked somewhere to the left of the Sea Demon.
5. We also know the Night Watcher is docked somewhere to the left of the ship loaded with emeralds (but we don't know her name).
6. The pirate with a wooden leg owns the ship that is docked somewhere to the left of the ship loaded with rubies.
7. There are two docks between the one where the ship loaded with coins is docked and the one where the ship belonging to the pirate with an eye patch is docked.
8. Also, there are two docks between the one where the Winter Fog is docked and the one where the ship belonging to the pirate with an eye patch is docked.
9. There are two docks between the one where the ship loaded with diamonds is docked and the one where the ship loaded with emeralds is docked.
10. The ship loaded with rubies is docked somewhere to the left of the one loaded with pearls.
11. There are two docks between the one where the Winter Fog is docked and the one where the Red Fury is docked, in some order.

The Docks at Pirate Bay

The Docks at Pirate Bay		Dock					Bounty					Feature				
		1	2	3	4	5	Coins	Diamonds	Emeralds	Pearls	Rubies	Eye patch	Face scar	Gold tooth	Nose earring	Wooden leg
Ship	Night Watcher															
	Red Fury															
	Sea Demon															
	Wave Crusher															
	Winter Fog															
Feature	Eye patch															
	Face scar															
	Gold tooth															
	Nose earring															
	Wooden leg															
Bounty	Coins															
	Diamonds															
	Emeralds															
	Pearls															
	Rubies															

Ship	Dock	Bounty	Feature

Anniversary Expeditions

These couples are celebrating their anniversaries in the coming months, and are all taking second honeymoons to celebrate. Each couple has their anniversary in a different month and is planning to visit a different destination.

Who is Augie's partner, what month is their anniversary, and where are they going to celebrate it?

1. Two things we know about Poppy.
   a. First, her anniversary is sometime later in the year than the anniversary of the couple visiting Turkey.
   b. Second, there is one month between Poppy's anniversary month and Zaid's anniversary month, in some order.
2. Speaking of the couple going to Turkey, their anniversary is sometime later in the year than the anniversary of the couple visiting Norway.
3. Joaquin's anniversary is sometime later in the year than Augie's anniversary.
4. Three things we know about Raiden.
   a. First, her anniversary is sometime later in the year than Zeke's anniversary.
   b. Second, her anniversary is either the month before or the month after Ariel's anniversary.
   c. Third, her anniversary is either the month before or the month after the anniversary of the couple traveling to Greece (but we know nothing about the couple going there).
5. Genevieve's anniversary is sometime later in the year than Tristan's anniversary.
6. Three things we know about the couple traveling to France.
   a. First, their anniversary is sometime earlier in the year than the anniversary of the couple traveling to Greece.
   b. Second, the couple visiting France celebrate their anniversary the month after Asha's anniversary.
   c. Third, there is one month between the anniversary month of the couple who will visit France and Tristan's anniversary month, in some order.

Anniversary Expeditions

Anniversary Expeditions	Month					Wife					Husband				
	January	February	March	April	May	Ariel	Asha	Genevieve	Poppy	Raiden	Augie	Joaquin	Tristan	Zaid	Zeke
Destination France															
Greece															
Italy															
Norway															
Turkey															
Husband Augie															
Joaquin															
Tristan															
Zaid															
Zeke															
Wife Ariel															
Asha															
Genevieve															
Poppy															
Raiden															

Destination	Month	Wife	Husband

The Room Service Riddle

The five guests occupying rooms 301 through 305 have ordered room service for breakfast. Each of them has ordered a different beverage, and each likes their eggs prepared a different way. Note their hotel rooms are in numerical order, with room 301 being the leftmost one.

Who ordered the orange juice, what kind of eggs went with it, and what room was this breakfast delivered to?

1. The person who ordered scrambled eggs is in a room somewhere to the left of the room that placed the order for parboiled eggs.
2. Two things we know about the person who ordered tea.
   a. First, there is one room between the tea-drinker's room and Cooper's room.
   b. Second, the tea drinker is in a room somewhere to the left of the room where the omelet will be delivered.
3. Two things we know about the person who ordered coffee.
   a. First, they're in a room somewhere right of the room belonging to the person who ordered an omelet.
   b. Second, there is one room between the coffee drinker's room and Ava's room.
4. Speaking of Ava, there is one room between hers and the room of the person who ordered fried eggs (whatever drink they ordered).
5. Three things we know about Nora.
   a. First, there is one room between her room and Ava's room.
   b. Second, there is one room between Nora's room and the water drinker's room (whose name we don't know).
   c. Third, her room is somewhere to the right of Ezequiel's room.
6. Cooper ordered hard boiled eggs.
7. There are two rooms between the room of the person who ordered an omelet and the room of the person who ordered milk.

The Room Service Riddle

The Room Service Riddle		Room					Eggs					Beverage				
		301	302	303	304	305	Fried	Hard boiled	Omelet	Parboiled	Scrambled	Coffee	Milk	Orange juice	Tea	Water
Name	Ava															
	Cooper															
	Ezequiel															
	Isaac															
	Nora															
Beverage	Coffee															
	Milk															
	Orange juice															
	Tea															
	Water															
Eggs	Fried															
	Hard boiled															
	Omelet															
	Parboiled															
	Scrambled															

Name	Room	Eggs	Beverage

Scavenger Hunt

Five friends are participating in a scavenger hunt they created for each other. Each of them took one small object and hid it in a different hiding place for the other four to find. The game was played over five rounds, with the player who hid the object refereeing that round of the hunt.

In which round did the players search for the ring, who was in charge of hiding it, and what hiding spot did they choose?

1. The player who used the bookshelf as a hiding spot hid their object sometime before the player who hid the die, but sometime after Shelby.
2. An object was hidden in a drawer either the round before or the round after Shelby hid her object.
3. There were also two rounds of play between the round where an object was hidden in a drawer and the round where a stamp was hidden, in some order.
4. There was one round between the round where an object was hidden in a pile of laundry and the round where an object was hidden under the rug, in some order.
5. Also, the object hidden under the rug was hidden sometime after Milo hid an object.
6. Simeon hid his object before Paco hid his, but didn't hide his object in the third round.
7. The object hidden under a pillow was placed there sometime before the coin was hidden.
8. The paper clip was hidden sometime after Harper hid an object.

Scavenger Hunt

	Scavenger Hunt	Round					Object					Hiding place				
		1	2	3	4	5	Coin	Die	Paper clip	Ring	Stamp	Bookshelf	Drawer	Laundry pile	Pillow	Under rug
Player	Harper															
	Milo															
	Paco															
	Shelby															
	Simeon															
Hiding place	Bookshelf															
	Drawer															
	Laundry pile															
	Pillow															
	Under rug															
Object	Coin															
	Die															
	Paper clip															
	Ring															
	Stamp															

Player	Round	Object	Hiding place

The Hidden Will

Mr. Dermott has purchased five little safes to hide his most valued possessions. Each one has a different numerical combination, he has put a different valuable item inside each of the safes, and each of them is hidden in a different room, in a specific location within said room.

Where is the will going to be hidden and what's the combination of the safe containing it?

1. Two things we know about the safe containing the passport.
    a. First, the first digit of its combination is somewhat higher than the first digit of the combination of the safe hidden under a floorboard, but lower than the first digits of the combinations of the safes in the ceiling and under the rug.
    b. Second, said first digit is somewhat higher than the first combination digit of the safe in the library (though we don't know where exactly the library safe is hidden).
2. The first digit of safe containing money is 2 higher than the first digit of the combination of the safe that's hidden in the dining room.
3. Three things we know about the safe containing jewelry.
    a. First, the first digit of its combination is 2 higher or 2 lower than the first digit of the combination of the safe that's hidden above a ceiling tile.
    b. Second, said first digit is somewhat higher than first digit of the combination of the safe that's hidden in the attic.
    c. Third, if we organize the combinations numerically, in ascending order, there is one combination number between the combination number of the safe containing jewelry and the combination number of the safe that's placed in the living room, in some order.
4. The first digit of the combination number of the safe containing Mr. Dermott's diary is 2 higher than the first digit of the combination of the safe that's hidden behind a painting. It's also somewhat lower than the first digit of the combination number of the safe that's hidden in the living room.

The Hidden Will

The Hidden Will		Combination					Room					Hiding place				
		1189	3204	5855	7900	9672	Attic	Bedroom	Dining room	Library	Living room	Above ceiling tile	Behind painting	Behind plant	Under floorboard	Under rug
Contents	Diary															
	Jewelry															
	Money															
	Passport															
	Will															
Hiding place	Above ceiling tile															
	Behind painting															
	Behind plant															
	Under floorboard															
	Under rug															
Room	Attic															
	Bedroom															
	Dining room															
	Library															
	Living room															

Contents	Combination	Room	Hiding place

The Magicians' Bunnies

Five different magicians are performing at birthday parties this week, one per day! Each magician has a bunny who performs along with them, and each bunny has a unique name and color.

Who is Fuzzy's owner, what is his fur color, and when will the duo perform?

1. There are two days between the day the tan bunny will perform with Callie and the day Blair will perform with her bunny, in some order.
2. Kaden and his bunny will perform sometime later in the week than Paige and her bunny.
3. Fuzzy and his owner will perform sometime earlier in the week than the cream bunny and its owner.
4. Three things we know about the black bunny.
   a. First, it will perform with its owner either the day before or the day after Cinnamon and her owner will perform.
   b. Second, the black bunny will perform with its owner sometime earlier in the week than Blair and her bunny.
   c. Third, the black bunny will perform with its owner either the day before or the day after the gray bunny and its owner perform (but we don't know the name of the gray bunny).
5. Four things we know about Twinkie.
   a. First, she is a fawn bunny.
   b. Second, her owner will perform with her sometime later in the week than the cream bunny and its owner will perform.
   c. Third, Twinkie and her owner will perform the day before Ajax and his owner.
   d. Fourth, Twinkie and her owner will perform later in the week than Kaden and his bunny (we don't have any additional information about the bunny, though).

The Magicians' Bunnies

The Magicians' Bunnies	Weekday					Bunny					Fur color				
	Monday	Tuesday	Wednesday	Thursday	Friday	Ajax	Cinnamon	Fluffy	Fuzzy	Twinkie	Black	Cream	Fawn	Gray	Tan
Magician Blair															
Callie															
Kaden															
Paige															
Peyton															
Fur color Black															
Cream															
Fawn															
Gray															
Tan															
Bunny Ajax															
Cinnamon															
Fluffy															
Fuzzy															
Twinkie															

Magician	Weekday	Bunny	Fur color

Capture the Flag

This weekend's Capture the Flag tournament was great! The five team leaders each created a different colored flag with a different mascot embroidered on it. In the end, each team received a ranking, 1-5 in the round robin tournament, with Rank 1 getting the grand prize.

What color team did Scarlett lead, what was the mascot of that team, and what ranking did that team receive?

1. Two things we know about the violet team.
   a. First, they placed above the team led by Elias in the tournament's final ranking.
   b. Second, one team placed between the violet team and the team with a bear on its flag, in some order.
2. We know two things about the team with the bear flag.
   a. First, that team and the team led by Eleanor placed next to each other, in some order.
   b. Second, the bear flag team earned a better final place in the tournament than the team with the shark flag.
3. Two things we know about the blue team.
   a. First, they earned a worse final place in the tournament than the team with the shark flag.
   b. Second, there are two teams that placed between the blue team and the green team, in some order.
4. Speaking of the green team, they earned a worse final placement than the team led by Santiago.
5. The team with a wolf on their flag earned a better final placement than the team led by Nova, but a worse placement than the yellow team.
6. The team with an eagle flag earned a better final placement than the team led by Eleanor.

Capture the Flag

Capture the Flag		Place					Mascot					Color				
		1	2	3	4	5	Bear	Eagle	Tiger	Shark	Wolf	Blue	Green	Red	Violet	Yellow
Leader	Eleanor															
	Elias															
	Nova															
	Santiago															
	Scarlett															
Color	Blue															
	Green															
	Red															
	Violet															
	Yellow															
Mascot	Bear															
	Eagle															
	Tiger															
	Shark															
	Wolf															

Leader	Place	Mascot	Color

97

The Magic Show Riddle

A great magician has come to town, for one week only! Each one of his five shows—one daily, Monday through Friday—will be unique and amazing. Each time he will perform a different opening act, have a different animal assistant, and close his show with daring escapism from a different great danger.

On what day will the magician escape being crushed, and what animal will accompany him onstage that day?

1. The magician will not open with mentalism on Monday.
2. Also, he will be assisted by a rabbit either the day before or the day after he does perform the mentalism trick.
3. Three things we know about the magician escaping a stabbing.
   a. First, there are two days between the day he escapes stabbing and the day he'll open with a coin trick, in some order.
   b. Second, there is one day between the day he'll attempt this escape and the day he'll open with a Rubik's cube trick, in some order.
   c. Third, he'll escape stabbing sometime after he performs a trick with a dog.
4. Two things we know about the magician attempting to escape incineration.
   a. First, he'll attempt this escape sometime before he performs a trick with a dog.
   b. Second, there is one day between the incineration escape and the day he'll escape beheading, in some order.
5. The magician will perform the Rubik's cube trick on a later weekday than the day he'll escape drowning.
6. The magician will perform a trick with a chick either the day before or the day after the show that opens with a pencil-through-glass trick, but he will definitely be performing with the chick sometime after he performs with a rabbit (whatever the opening act on that day is).
7. The tricks involving a duck and a rabbit will be performed on consecutive days, in some order.

98

The Magic Show Riddle

The Magic Show Riddle		Weekday					Animal					Danger				
		Monday	Tuesday	Wednesday	Thursday	Friday	Chick	Dog	Duck	Pigeon	Rabbit	Beheaded	Crushed	Drowned	Incinerated	Stabbed
Opening act	Card trick															
	Coin trick															
	Mentalism															
	Pencil-through-glass															
	Rubik's Cube trick															
Danger	Beheaded															
	Crushed															
	Drowned															
	Incinerated															
	Stabbed															
Animal	Chick															
	Dog															
	Duck															
	Pigeon															
	Rabbit															

Opening act	Weekday	Animal	Danger

The Butterfly Collectors

Five friends are avid butterfly collectors. This week they were extremely lucky: each of them captured a different colored butterfly, in a different location, on a different day of the week.

Who captured the blue butterfly, and where did they capture it?

1. The white butterfly was captured sometime after the butterfly caught in the countryside.
2. Allie captured a butterfly sometime after a butterfly was caught in the street.
3. Two things we know about Sam.
   a. First, there is one day between the day he captured a butterfly and the day the green butterfly was captured, in some order.
   b. Second, he captured a butterfly either the day before or the day after someone captured a butterfly in the street.
4. Peter captured a butterfly sometime after a butterfly was caught at home.
5. He also caught his butterfly either the day before or the day after somebody caught a butterfly in the forest.
6. There is one day between the day someone caught a butterfly at their home and the day someone caught a butterfly in a park, in some order.
7. Three things we know about the purple butterfly.
   a. First, it was captured sometime after the butterfly in the countryside was caught.
   b. Second, it was captured either the day after or the day before Allie caught a butterfly, but definitely sometime before Richie caught his butterfly.
   c. Third, there are two days between the day the purple butterfly was caught and the day the orange butterfly was caught, in some order.

The Butterfly Collectors

The Butterfly Collectors	Weekday					Location					Color				
	Monday	Tuesday	Wednesday	Thursday	Friday	Countryside	Forest	Home	Park	Street	Blue	Green	Orange	Purple	White
Name Allie															
Paul															
Peter															
Richie															
Sam															
Color Blue															
Green															
Orange															
Purple															
White															
Location Countryside															
Forest															
Home															
Park															
Street															

Name	Weekday	Location	Color

Great Families of Old

Spencer is writing a saga about five noble families in medieval times. Each one of his five books features one of these families.

What is the motto and coat of arms of the family in the fourth book?

1. Two things we know about the Jilton family.
    a. First, they appear in a later book than the one featuring the coat of arms with the flower, but in an earlier book than the coat of arms with the eagle.
    b. Second, there is one book between the one featuring the Jilton family and the one featuring the coat of arms with the wolf, in some order.
2. Two things we know about the coat of arms with the lion.
    a. First, it appears in the book just before or just after the book featuring the motto "Strength and Power."
    b. Second, there is one book between the one featuring this coat of arms and the one with the dragon coat of arms, with the lion coat of arms appearing earlier.
3. The family whose coat of arms has an eagle appears in a book released sometime after the book featuring the family with the motto "Always Better."
4. The family with the motto "Live with Pride" appears in a book released sometime before the one featuring the motto "Always Better," but sometime after the one featuring the Garrett family.
5. Two things we know about the August family.
    a. First, there are two books between the one featuring them and the one featuring the Ervine family, in some order.
    b. Second, the August family features in a book released sometime before the book featuring the motto "Face the Storm."

Great Families of Old

	Great Families of Old	Book					Motto					Coat				
		1	2	3	4	5	Always Better	Face the Storm	Guided by...	Live with Pride	Strength and...	Dragon	Eagle	Flower	Lion	Wolf
Family	August															
	Ervine															
	Garrett															
	Jilton															
	Lambert															
Coat	Dragon															
	Eagle															
	Flower															
	Lion															
	Wolf															
Motto	Always Better															
	Face the Storm															
	Guided by...															
	Live with Pride															
	Strength and...															

Family	Book	Motto	Coat

Socialite Sightings

A celebrated socialite is in town. Each day of last week, she visited a different part of town, wearing a different dress, and a necklace with a different gemstone.

What necklace and what color dress was she wearing when she was spotted in the museum?

1. The socialite went to the public library either the day before or the day after she wore the necklace with a jade gemstone.
2. She also wore the jade necklace some day after she wore the blue dress.
3. The lady wore the necklace with an amethyst some day after she wore the necklace with an opal.
4. Also, she wore the necklace with an opal either the day before or the day after she wore the black dress.
5. She certainly didn't wear the necklace with an amethyst on Friday.
6. The lady wore the purple dress some day after she wore the necklace with an onyx gemstone, but some day before she went to the park.
7. The blue dress was also worn some day after the necklace with an onyx gemstone.
8. There was one day between the day the lady was seen in the restaurant and the day she wore a necklace with an opal, in some order.
9. The lady went to the park some day before she went to the theater.
10. The lady wore the black dress some day after she wore the mauve one.

104

Socialite Sightings

	Weekday					Necklace					Dress color				
Socialite Sightings	Monday	Tuesday	Wednesday	Thursday	Friday	Amethyst	Jade	Onyx	Opal	Turquoise	Black	Blue	Mauve	Purple	Red
Location Museum															
Park															
Public library															
Restaurant															
Theater															
Dress color Black															
Blue															
Mauve															
Purple															
Red															
Necklace Amethyst															
Jade															
Onyx															
Opal															
Turquoise															

Location	Weekday	Necklace	Dress color

The Power Stones

The five power stones must be assembled! We need your help to locate each one, determine what characteristic differentiates each from the rest, and the correct order to use them so their full potential is unleashed? We must determine these for each stone to collect them in one place and save the world!

What's the characteristic of the earth stone, and where would you travel to find it?

1. Three things we know about the stone you can find in the swamp.
    a. First, it must be used either just before or just after the transparent stone.
    b. Second, you must use the stone from the swamp at some point after the stone that's always warm.
    c. Third, it must be used just before or just after the wind stone (though we don't know anything about the characteristic of the wind stone).
2. The transparent stone must be used at some point before the heavy stone, and also at some point before the wind stone (though we still don't know the wind stone's characteristic).
3. Two things we know about the fire stone.
    a. First, you must use one stone between the fire stone and the stone found in the desert, in some order.
    b. Second, you must use the fire stone at some point before the water stone.
4. Two things we know about the rough stone.
    a. First, it must be used either just before or just after the water stone.
    b. Second, you must use one stone between the rough stone and the stone in the cave, in some order.
5. Two things we know about the warm stone.
    a. First, you must use one stone between the warm stone and the heart stone, in some order.
    b. Second, you must use the warm stone either just before or just after the stone in the forest.

106

The Power Stones

The Power Stones		Order					Location					Characteristic				
		1	2	3	4	5	Cave	Desert	Forest	Mountain	Swamp	Glows	Heavy	Rough	Transparent	Warm
Stone	Earth															
	Fire															
	Heart															
	Water															
	Wind															
Characteristic	Glows															
	Heavy															
	Rough															
	Transparent															
	Warm															
Location	Cave															
	Desert															
	Forest															
	Mountain															
	Swamp															

Stone	Order	Location	Characteristic

Cookies With Grandma

There's nothing better than visiting grandma! Grandma Gretel loves cookies, and so do her grandchildren. This week, each of her five grandchildren will bake cookies with Grandma Gretel on a different day, with different shapes and flavors being baked each day.

What flavor cookies will Maverick bake with grandma, and what shape will those delicious cookies be?

1. Three things we know about the lemon cookies.
   a. First there are two days between the day when they'll be baked and the day the heart cookies will be baked, in some order.
   b. Second, they will be baked at some point after Gilbert bakes with grandma.
   c. Third, they will be baked sometime before the peanut butter cookies, but after the star-shaped cookies.
2. We know two things about Hazel.
   a. First, there is one day between the day when she will bake with grandma and the day Autumn will bake with grandma, in some order.
   b. Second, there is one day between the day Hazel will bake with grandma and the day the dog-shaped cookies will be baked, in some order (although we don't know who will bake these).
3. Three things we know about the cat-shaped cookies.
   a. First, there is one day between the day the cat-shaped cookies will be baked and the day grandma will bake cookies with Lucille, in some order.
   b. Second, they'll be baked either the day before or the day after the chocolate chip cookies are baked.
   c. Third, they will be baked at some point after the oatmeal raisin cookies are baked.
4. There are two days between the day the sugar cookies will be baked and the day the moon-shaped cookies will be baked, with the sugar cookies being made first.

Cookies With Grandma

Cookies With Grandma		Weekday					Flavor					Shape				
		Monday	Tuesday	Wednesday	Thursday	Friday	Chocolate chip	Lemon	Oatmeal raisin	Peanut butter	Sugar	Cat	Dog	Heart	Moon	Star
Grandchild	Autumn															
	Gilbert															
	Hazel															
	Lucille															
	Maverick															
Shape	Cat															
	Dog															
	Heart															
	Moon															
	Star															
Flavor	Chocolate chip															
	Lemon															
	Oatmeal raisin															
	Peanut butter															
	Sugar															

Grandchild	Weekday	Flavor	Shape

Know Your Neighbors

You've just moved into a new neighborhood, on the opposite side of Summerfield Street from Houses 621 through 625. 621 is the leftmost house of the row, and each house is home to a different person with a different job and a different pet.

What is the name, occupation, and pet of the person living in house 624?

1. Two things we know about the dog.
    a. First, there is one house between the dog's house and Artemis' home.
    b. Second, there are two houses between the dog's house and the house where the writer lives.
2. Two things we know about Everley.
    a. First, she lives somewhere left of the doctor.
    b. Second, there are two houses between Everly's house and the parrot owner's house, in some order.
3. Two things we know about the teacher.
    a. First, there is one house between the teacher's home and Seraphina's home.
    b. Second, there are two houses between the teacher's house and the house with the fish.
4. There are two houses between the actor's house and the writer's house.
5. Two things we know about Adrian.
    a. First, there is one house between his house and that of the cat owner's house.
    b. Second, he lives somewhere left of the teacher.

Know Your Neighbors

Know Your Neighbors		House					Occupation					Pet				
		621	622	623	624	625	Actor	Doctor	Lawyer	Teacher	Writer	Bunny	Cat	Dog	Fish	Parrot
Name	Adrian															
	Artemis															
	Everley															
	Lia															
	Seraphina															
Pet	Bunny															
	Cat															
	Dog															
	Fish															
	Parrot															
Occupation	Actor															
	Doctor															
	Lawyer															
	Teacher															
	Writer															

Name	House	Occupation	Pet

Solutions

Five Ships

Name	Dock	Captain	Merchandise
Fairy	1	Lennox	Gold
Mermaid	3	Murdoch	Jewels
Neptune	5	Croydon	Silk
Sea Queen	2	Harris	Spices
Valentine	4	Millman	Tobacco

Restaurant Recommendations

Name	Stars	Cuisine	Color
Delicious	3	Italian	Yellow
Happy Tummy	3.5	Indian	Red
Hungry No More	4.5	French	Gray
Nice Food Diner	4	Chinese	White
Yum Yum	5	Thai	Blue

The Christmas Ornaments Riddle

Child	Date	Ornament	Color
Avery	19	Reindeer	White
Elijah	22	Santa	Gold
Mateo	21	Star	Silver
Paisley	20	Angel	Green
Stella	23	Bell	Red

Custom Dice

Feline	Order	Shape	Letter
Cheetah	3	Square	H
Leopard	2	Hexagon	D
Lion	5	Triangle	F
Panther	4	Pentagon	Z
Tiger	1	Circle	A

Alien Planets

Name	Parsecs	Personality	Sun color
Ardyk	10	Aggressive	Red
Dongyhuk	30	Shy	Gray
Gamush	20	Selfish	Blue
Reshyu	40	Suspicious	Purple
Thaket	50	Friendly	Green

Forest Findings

Contents	Box	Wood	Key color
Feathers	3	Ash	Lilac
Flower petals	4	Alder	Mint
Leaves	1	Cherry	Silver
Pine cones	2	Hickory	Tan
Stones	5	Aspen	Gold

A Week's Worth of Meals

Sandwich	Day	Dessert	Juice
Ham	Monday	Cookie	Orange
Cheese	Tuesday	Cupcake	Peach
Egg salad	Wednesday	Granola bar	Pineapple
Tomato	Thursday	Yogurt	Lemonade
Tuna	Friday	Fruit	Apple

Book Day

Name	Desk	Book title	Genre
Arnie	1	The Best of Jokes	Action
Cam	5	Playing the Odds	Sci-Fi
Ellie	3	The Last Day...	Horror
Julie	2	Finding Theo	Graphic novel
Vince	4	For All the...	Detective

All the Luck They Can Get

Student	Weekday	Lucky charm	Test subject
Emma	Friday	Die	Literature
Gail	Tuesday	Horseshoe	English
Jay	Thursday	Jacket	Biology
Mabel	Wednesday	Coin	History
Norm	Monday	Rabbit's paw	Math

A Five-Step Password

Shape	Step	Color	Word
Circle	4	Yellow	Cat
Moon	1	Red	Bird
Square	3	Green	Dog
Star	2	Purple	Fish
Triangle	5	Blue	Rabbit

A Week at the Colosseum

Gladiator	Weekday	Weapon	Animal
Caius	Wednesday	Sword	Wolf
Cassius	Tuesday	Lance	Tiger
Marcellus	Friday	Sling	Lion
Rufus	Monday	Bow and arrows	Bear
Theodorus	Thursday	Club	Dog

Pizza With Everything!

Meat	Table	Soda	Non-meat
Bacon	4	Root beer	Tomato
Chicken	2	Orange	Onions
Ham	1	Lemon-lime	Olives
Pepperoni	5	Cola	Paprika
Sausage	3	Ginger ale	Mushrooms

Avian Artistry

Bird	Month	Technique	Dedicated to
Eagle	August	Sketch	Brother
Owl	April	Oil painting	Mom
Raven	May	Ink painting	Dad
Sparrow	June	Photograph	Sister
Woodpecker	July	Cartoon	Friend

The Five Temples

Master	Visit number	Spirit	Lesson
Gonji	1	Eagle	Strength
Josun	3	Monkey	Wisdom
Pahuk	4	Dragon	Bravery
Raku	5	Crane	Patience
Teshi	2	Phoenix	Generosity

Strange Locker Combinations

Number	Locker	Shape	Letter
29	2	Circle	Y
37	1	Square	Z
60	4	Rhombus	X
82	5	Triangle	W
94	3	Rectangle	Q

Sacrifices to the Gods

Name	Weekday	Temple	Offering
Constantine	Friday	Apollo	Goat
Daphne	Thursday	Poseidon	Bull
Phoebe	Wednesday	Artemis	Sheep
Theodore	Monday	Athena	Goose
Zoe	Tuesday	Zeus	Horse

Driver App Evals

Name	Rating	Plate	Car
Betty	4.7	PJZ458	Mitsubishi
Casper	5.0	HYU792	Citröen
Darius	4.8	QSD580	Chevrolet
Lydia	4.6	JKR143	Subaru
Ronnie	4.9	TKI327	Peugeot

The Rose Contest

Owner	Place	Species	Color
Mrs. Dhalmar	3	Grandiflora	Pink
Mrs. Griffin	5	Floribunda	Orange
Mrs. Morris	1	Hybrid Musk	White
Mrs. Potter	2	Hybrid Tea	Yellow
Mrs. Thompson	4	Damask	Red

Slot Machine Madness

Shape	Number	Fruit	Animal
Hearts	5	Lemons	Fish
Moons	1	Oranges	Cats
Stars	4	Cherries	Dogs
Suns	2	Grapes	Mice
Triangles	3	Apples	Birds

Cellblock Countdown

Hair color	Prison cell	Scar location	Eye color
Black	1	Arm	Green
Blond	3	Hand	Amber
Brown	5	Face	Gray
Gray	2	Leg	Hazel
Red	4	Neck	Blue

Furniture Purchases

Family	Size	Model	Type
Abbott	S	Alexia	Chair
Brenton	L	Jasper	Armchair
Clarke	XL	Roman	Sofa
Jackson	M	Halley	Bed
McQueen	XS	Rubicon	Table

Consulting the Gods

Name	Weekday	Topic	Method
Attilius	Wednesday	Wedding	Weather patterns
Julius	Monday	Sailing	Bird flight
Octavius	Thursday	Politics	Oracle
Marcus	Tuesday	War	Chicken feed
Quintus	Friday	Business	Animal entrails

The Sticker Collection

Friend	Weekday	Person	Color
Cassandra	Tuesday	Henry VIII	Orange
Claudius	Wednesday	Joan of Arc	White
Hamlet	Thursday	Julius Caesar	Pink
Ophelia	Friday	Empress Elizabeth	Gold
Titus	Monday	Napoleon	Red

Boggling Backpacks

Child	Seat	Color	Design
Brenda	5	Blue	Dog
Eddie	2	Brown	Lion
Gary	1	Orange	Dinosaur
Kelly	3	Purple	Bear
Vanessa	4	Yellow	Tiger

Summer Fun

Planner	Weekday	Game	Sport
Dad	Friday	Dominoes	Hiking
Daughter	Wednesday	Liar's Dice	Basketball
Grandmother	Tuesday	Monopoly	Football
Mom	Thursday	Blackjack	Biking
Son	Monday	Scrabble	Swimming

The Tattoo Artist

Client	Weekday	Design	Location
Ava	Tuesday	Yin-yang	Leg
Chuck	Monday	Dragon	Arm
Dave	Wednesday	Skull	Neck
Garrett	Thursday	Rose	Shoulder
Sue	Friday	Name	Chest

The Journey Through Town

Street	Step	Shop	Shopkeep
Acorn	3	Bakery	Theo
Frazer	1	Inn	Adam
Gladwell	5	Barber	Claudia
Millboy	4	Butcher	Brian
Sweeney	2	Jeweler	Angela

A Rainy Week

Host	Weekday	Drink	Food
Kinsley	Friday	Coffee	Apple pie
Leilani	Tuesday	Hot milk	Orange cake
Parker	Monday	Tea	Cookies
Rowan	Wednesday	Hot cocoa	Waffles
Sawyer	Thursday	Spiced wine	Brownies

Card Tricks

Card chosen	Weekday	Effect	Source
Ace of Spades	Monday	On top of deck	YouTube
Three of Spades	Wednesday	Appears in hand	Friend
Seven of Hearts	Tuesday	Face down in deck	Book
Nine of Diamonds	Friday	Appears in pocket	Magic lesson
King of Clubs	Thursday	Disappears from deck	Dad

Start Your Engines

Driver	Lane	Car	National
Bernard	3	White	USA
Carl	2	Purple	Germany
Dwayne	4	Red	Canada
Sheldon	5	Blue	Denmark
Toby	1	Green	Brazil

The Aquarium Dolphins Riddle

Name	Weekday	Trick	Treat
Captain	Thursday	Lift trainer	Squid
Flipper	Monday	Tail wave	Salmon
Rocky	Friday	Bounce ball	Shrimp
Shamu	Tuesday	High five	Mussels
Wally	Wednesday	Hoop jump	Tuna

The Birthday Celebrations Riddle

Lunch	Age	Activity	Present
Burgers	11	Movie	Book
Pizza	13	Pool party	Toy robot
Chicken	14	Theater	Art set
Sushi	12	Museum	Jigsaw puzzle
Tacos	15	Soccer	Soccer ball

Dragons and Gemstones

Name	Cave	Gemstone	Color
Fryn	3	Sapphire	White
Garwyn	5	Amethyst	Red
Khamu	4	Diamond	Blue
Nelbon	1	Ruby	Green
Tipur	2	Emerald	Black

Five Chess Sets

Set style	Month	Light pieces	Dark pieces
Dragon	October	Pink	Blue
Elf	June	Yellow	Violet
Fairy	July	Beige	Indigo
Wolf	August	Orange	Green
Wizard	September	Red	Brown

Cozy Caves

Rug skin	Cave	Rock painting	Food
Cow	5	Ritual	Nuts
Deer	4	Night sky	Meat
Hare	1	Family	Roots
Horse	2	Lake	Vegetables
Mammoth	3	Hunt	Fruit

The Five Newborns

Name	Cradle	Hair color	Onesie pattern
Arlo	2	Red	Clowns
Caspian	1	Bald	Rockets
Isla	3	Black	Teddy bears
Maeve	5	Blonde	Animals
Silas	4	Brown	Trains

The Docks at Pirate Bay

Ship	Dock	Bounty	Feature
Night Watcher	1	Diamonds	Wooden leg
Red Fury	5	Pearls	Eye patch
Sea Demon	3	Rubies	Face scar
Wave Crusher	4	Emeralds	Gold tooth
Winter Fog	2	Coins	Nose earring

Anniversary Expeditions

Destination	Month	Wife	Husband
France	February	Ariel	Zaid
Greece	April	Poppy	Tristan
Italy	May	Genevieve	Joaquin
Norway	January	Asha	Zeke
Turkey	March	Raiden	Augie

The Room Service Riddle

Name	Room	Eggs	Beverage
Ava	302	Omelet	Water
Cooper	303	Hard boiled	Orange juice
Ezequiel	301	Scrambled	Tea
Isaac	305	Parboiled	Milk
Nora	304	Fried	Coffee

Scavenger Hunt

Player	Round	Object	Hiding place
Harper	1	Ring	Drawer
Milo	3	Coin	Laundry pile
Paco	5	Die	Under rug
Shelby	2	Paper clip	Pillow
Simeon	4	Stamp	Bookshelf

The Hidden Will

Contents	Combination	Room	Hiding place
Diary	7900	Dining room	Above ceiling tile
Jewelry	5855	Bedroom	Behind painting
Money	9672	Living room	Under rug
Passport	3204	Attic	Behind plant
Will	1189	Library	Under floorboard

The Magicians' Bunnies

Magician	Weekday	Bunny	Fur color
Blair	Tuesday	Cinnamon	Gray
Callie	Friday	Ajax	Tan
Kaden	Wednesday	Fluffy	Cream
Paige	Monday	Fuzzy	Black
Peyton	Thursday	Twinkie	Fawn

Capture the Flag

Leader	Place	Mascot	Color
Eleanor	3	Wolf	Red
Elias	5	Tiger	Blue
Nova	4	Shark	Violet
Santiago	1	Eagle	Yellow
Scarlett	2	Bear	Green

The Magic Show Riddle

Opening act	Weekday	Animal	Danger
Card trick	Monday	Pigeon	Incinerated
Coin trick	Tuesday	Dog	Drowned
Mentalism	Friday	Chick	Stabbed
Pencil-through-glass	Thursday	Rabbit	Crushed
Rubik's Cube trick	Wednesday	Duck	Beheaded

The Butterfly Collectors

Name	Weekday	Location	Color
Allie	Wednesday	Home	Green
Paul	Tuesday	Street	Purple
Peter	Friday	Park	Orange
Richie	Thursday	Forest	White
Sam	Monday	Countryside	Blue

Great Families of Old

Family	Book	Motto	Coat
August	2	Live with Pride	Flower
Ervine	5	Face the Storm	Dragon
Garrett	1	Guided by...	Wolf
Jilton	3	Always Better	Lion
Lambert	4	Strength and...	Eagle

Socialite Sightings

Location	Weekday	Necklace	Dress color
Museum	Tuesday	Turquoise	Purple
Park	Wednesday	Opal	Blue
Public library	Thursday	Amethyst	Black
Restaurant	Monday	Onyx	Mauve
Theater	Friday	Jade	Red

The Power Stones

Stone	Order	Location	Characteristic
Earth	2	Cave	Warm
Fire	1	Forest	Glows
Heart	4	Swamp	Rough
Water	3	Desert	Transparent
Wind	5	Mountain	Heavy

Cookies With Grandma

Grandchild	Weekday	Flavor	Shape
Autumn	Wednesday	Chocolate chip	Dog
Gilbert	Monday	Oatmeal raisin	Heart
Hazel	Friday	Peanut butter	Moon
Lucille	Tuesday	Sugar	Star
Maverick	Thursday	Lemon	Cat

Know Your Neighbors

Name	House	Occupation	Pet
Adrian	623	Doctor	Bunny
Artemis	624	Teacher	Parrot
Everley	621	Lawyer	Fish
Lia	625	Writer	Cat
Seraphina	622	Actor	Dog

Thank you for reading!

We hope this book was useful for you, and we look forward to seeing you in our published puzzle books!

Made in the USA
Columbia, SC
16 June 2025

59445881R00072